Concrete afloat

Proceedings of the conference on concrete ships and floating structures organized by The Concrete Society in association with the Royal Institution of Naval Architects and held in London on 3 and 4 March, 1977

Thomas Telford Ltd
London, 1977

Published by Thomas Telford Ltd, Telford House, PO Box 101,
26–34 Old St, London EC1P 1JH

ISBN: 0 7277 0048 0

Text set in 11/12 pt Photon Times, printed by photolithography, and bound in Great Britain at The Pitman Press, Bath

Contents

CONVERSION FACTORS

Length	1 in. 1 ft 1 mile	0·0254 m 0·3048 m 1609·344 m
Area	1 ft^2	·092 903 0 m^2
Volume	1 ft^3 1 US barrel (petroleum)	0·028 316 8 m^3 158·987 dm^3
Velocity	1 international knot 1 UK knot	0·514 444 m/s 0·514 773 m/s
Mass	1 UK ton 1 short ton	1·016 05 t 0·907 185 t
Density	1 lb/ft^3	16·0185 kg/m^3
Volume rate of flow	1 UK gal/h	1·262 80 × 10^{-6} m^3/s
Force	1 tonne-force 1 kp	9·806 65 kN 9·806 65 N
Pressure	1 lbf/in^2	6·894 76 kPa
Energy	1 Btu	1055·06 J
Power	1 hp	745·700 W

Concrete afloat. Thomas Telford Ltd, London, 1977, 1–21

The paper traces the development of ferrocement and reinforced and prestressed concrete as hull materials for a wide range of floating structures. The distinct characteristics of shipbuilding concrete in terms of material properties, design, construction techniques, hull shape, size and weight are critically examined in relation to cargoes carried, special uses and short and long term operational behaviour.

Development of the concrete hull

R. G. MORGAN, BSc, MICE*

'When a new idea or invention is first put into service it is naturally received with a certain amount of antagonism; the concrete ship is no exception to this rule. The doubts still abounding in regard to the concrete ship arise principally from ignorance of what has already been accomplished; and in order to appreciate what has been accomplished, and so acquire a basis for the formation of an intelligent opinion on the subject of concrete ships, it is desirable to study, however briefly, the development in the use of other shipbuilding materials.'

This is the opening paragraph of N. K. Fougner's book[1] written in 1922, in which he describes the design, construction and operation of the many concrete ships built during the World War I period. Fougner was writing about ships built using conventional reinforced concrete only but his comments are equally true today when ferrocement and prestressed concrete have taken over from reinforced concrete as the principal materials for hull construction.

The development of the concrete hull can be divided into four phases (see Fig. 1): the two major shipbuilding programmes of the World War periods when the depletion of the merchant ships of the combating nations forced governments to use concrete to save the steel needed for military purposes, and the two 'peace-time' phases from about 1900 to 1916 and from 1960 to the present day when concrete must compete on equal terms with other shipbuilding materials in terms of building costs, strength, safety, durability, operational qualities and overall economic viability.

The mention of concrete as a material for shipbuilding still brings the sceptical comments that it is too heavy and will not float and that it is so brittle that it will break into pieces on impact—the very same objections raised when iron was introduced at the beginning of the nineteenth century. Concrete also suffers from the limited experience of too many people who see only cracked paving slabs and rusting fence posts, so concrete needs some introduction.

* Lecturer in Civil Engineering, University of Bristol.

CONCRETE

Concrete is a man-made material produced from a mixture of cement, fine and coarse aggregates and water which in its plastic state has to be formed into the required shape and on hardening develops high compressive strength and possesses excellent durability.

The constituents and their proportions are selected to produce a range of concretes (with aggregates bigger than 5 mm) or mortars (no aggregates larger than 5 mm) varying in workability or plasticity, depending on whether the material is to be cast inside moulds or plastered or sprayed. The chemical reaction between cement and water which binds the aggregate together gives an increasing strength at a decreasing rate over several years. In the first few days the concrete needs to be maintained in a warm and humid environment so that it may develop its strength properly—this is called curing—and also it needs protection when it is still comparatively weak as early shrinkage due to loss of water would cause surface crazing.

Concrete will expand and contract with moisture and temperature changes. The strain when first loaded is followed by a further deformation (called creep) under constant load which progresses at a diminishing rate.

Concrete is a complex material and it is important therefore that its production, application and curing are properly controlled.[2] Concrete

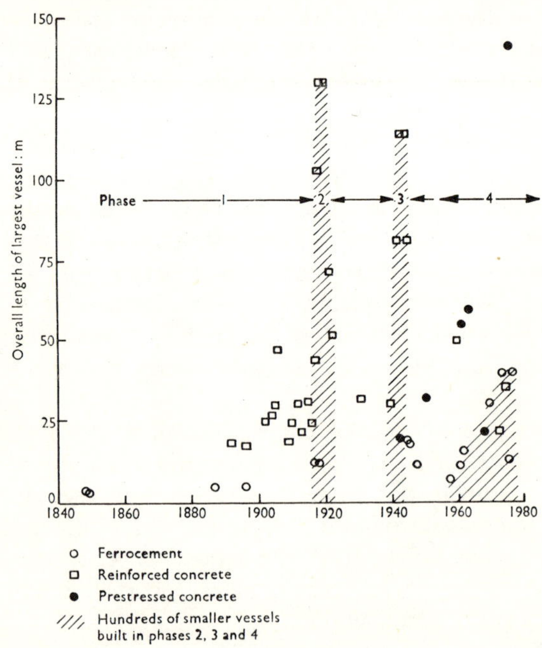

Fig. 1. Chart showing largest vessel built in any one year, 1848–1977

though strong in compression cracks under low tensile loading and the quantity, size and dispersion of steel bars and meshes are carefully chosen to carry the tensile stresses in the concrete due to bending, shear and torsion and to ensure that cracks formed under load are kept to acceptable limits so that no corrosion of the reinforcement takes place. Alternatively, prestressing techniques can be used to induce a compressive stress in the concrete so that under operational loads a favourable and usually crack-free stress condition is present.

Because of its high durability, concrete (using natural cements) had been used by the Romans and Ancient Egyptians for harbours and breakwaters, and Smeaton developed Portland cement as a jointing material for the Eddystone lighthouse in 1756. In 1848 and 1849 Joseph Louis Lambot, a French horticulturist, built two small rowing boats by plastering a sand–cement mortar over a framework of iron bars and mesh and both boats have survived[3,4] (Fig. 2).

Fig. 2. Lambot's boat of 1849, in Paris in 1974

Fig. 3. Gabellini's boat, 1896

Fig. 4. Modern ferrocement lay-up

FERROCEMENT

Gabellini[5] and Boon[6] and many others followed Lambot's intuitive concept but it was not until 1943 that Nervi[7] carried out the first recorded experiments on ferrocement and sought to understand its properties and structural behaviour before building several vessels immediately after the war. A few research papers were published in the early 1950s and a number of boats were built in Australia, New Zealand and the UK in 1960. Interest was then aroused in ferrocement and thousands of vessels have been built ranging from the 7·3 m long, 10 mm thick planing hull capable of 56 km/hour[8] to 50 m long fishing vessels 40 mm thick. Research and development in design and construction is catching up with and influencing practice and a wide variety of papers have been published at several conferences.[9–11] The most comprehensive research programme has been conducted at British Columbia Research at Vancouver[12] for the Canadian Government, admirably summarized by Greenius[13] in 1975. A list of research projects has been given by Morgan,[14,15] and Bowen and Baugh[16] have highlighted future research needs. Several design papers have been presented at conferences[10] and also by Fyson,[17] Bigg[18] and Johansen.[19]

Lambot used a framework of iron bars to give the basic hull shape before covering it with one layer of mesh, while Nervi dispensed with the heavier bars and relied on several layers of mesh to give the required strength. Both these techniques are still used today—compare Gabellini's boat of 1896 (Fig. 3) with a modern pipe-framed lay-up (Fig. 4). As vessels get bigger, steel web lattices are often used as transverse frames, and high tensile bars run longitudinally and transversely to give overall strength; the major roles of the mesh are to hold the mortar in place during application and finely disperse the cracks that develop under tensile loading. Alternatively, male or female moulds shape the reinforcement, and the mortar may be gunited (or sprayed).[20] As mass production methods develop, completely moulded systems, already used extensively in China, will prove more economical for identical vessels and cut out much of the highly labour-intensive operation.

The early ferrocement builders did not know that increasing the water content in their mortars would dramatically reduce both strength and durability, but fortunately only a small amount of water (usually measured as water/cement ratio by weight and being about 0·4 for a mix of one part of concrete to two parts of sand) was needed to make the mortar easily worked into the meshes. They did, however, recognize the need for full penetration and compaction and kept the sand size below about 3 mm.

Ferrocement should not be mistaken as a form of micro-reinforced concrete, for it has a very much higher proportion of steel dispersed throughout the section and this is usually measured in terms of the specific surface, k, defined as the surface area of all reinforcement per unit volume of the composite. A minimum requirement for ferrocement is usually taken as $0{\cdot}2\ mm^{-1}$ with a mortar cover of 2 mm in a thin section, whereas a k value of less than $0{\cdot}05\ mm^{-1}$ would suggest reinforced concrete.[21]

REINFORCED CONCRETE

Conventional reinforced concrete means the use of bar reinforcement, 12–32 mm dia. in hull thicknesses of 100 mm or more, placed inside formwork or moulds; and concrete, with large aggregates 5–25 mm as well as sand, poured and compacted inside. The large cover to fewer widely spaced bars gives less control over cracking, and critical crack widths of 0·05 mm occur at lower service loads than for ferrocement; reinforced concrete can only be used for marine structures where loading stresses are low, otherwise large cracks will cause the reinforcement to corrode.

For both ferrocement and reinforced concrete the ultimate direct tensile strength is equal to the strength of the reinforcement in the direction of load, but the strength in service is limited by the development of cracks of critical width. Similarly in flexure, the crack pattern and widths can be calculated and crack sizes are governed primarily by the spacing of mesh or reinforcing bars nearest the surface. In compression the load is shared by the steel and concrete, due account being taken of the effective stiffness of the hull. Classical reinforced concrete theory can be applied to ferrocement; however, design manuals and codes of practice for reinforced concrete are based on the needs for and experience from land-based structures, and marine structures need more refined analysis and details.

The change to reinforced concrete was made for the construction of pontoons and barges at the turn of the century: Searle[22] refers to a pontoon built in Rome in 1897 and both he and Højgaard[23] credit Gabellini with the first concrete barge, a 50 t lighter built in 1902. Gabellini built many barges, including the 150 t *Liguria* built in 1905 and reported as giving good service twelve years later. The first British vessel *Sand witch*[24] was built in 1910 and the Manchester Ship Canal pontoon in 1912.[25] Searle[22] refers to very many pontoons and barges built all over the world in the period leading up to World War I.

The Norwegian engineer N. K. Fougner[1] was the first to build a seagoing vessel: his pioneer *Namsenfjord* (182 t dead weight, 25·5 m long) was launched in 1917 as an experiment based on his experience in building concrete lighters. An ocean-going ship needed government approval and after much debate permission was finally given. Fougner chose to pour his hull walls between supported and braced metal laths; the two surfaces were afterwards finished with a cement mortar, thus

Fig. 5. Faith leaving Vancouver harbour

creating a ferrocement outer layer.

World War I caused major losses in merchant shipping due to submarine action, and plate steel was needed for military use. The UK and the USA turned to reinforced concrete as an alternative hull material and emergency shipbuilding programmes were established. The development, design, construction and operational qualities of many large vessels built in several countries during the war period are described by Højgaard,[23] Fougner,[1] Finsterwalder[26] and Haviland.[27] Most of the big ships came from the USA, including the biggest ever built, *Selma* (6340 t, 130 m long), and one of the most successful, the privately built *Faith* (4185 t, 97 m long) (Fig. 5). In the UK a few ocean-going vessels were built, [28–30] and *Armistice* (1150 t, 62·5 m long) was afloat and in use until a few years ago.[31] Fig. 6 shows a cross section of *Armistice* typical of framed reinforced concrete vessels of both World War periods. Shipyards were set up in many parts of the UK to build 1000 t capacity barges,[32] all named 'Crete-something'; one such barge, *Creteravine*, built in 1919 in Gloucester, is still afloat in a Norwegian fjord. About 85 000 t of seagoing shipping was built in this period, not counting several hundred barges, lighters and pontoons and a few floating docks.

A glut of merchant shipping followed the end of hostilities; and concrete shipbuilding programmes eventually stopped and no new concrete vessels were built after 1922.

Nothing was done to develop concrete as a hull material, and the whole process was repeated in World War II, with by far the bulk of shipping being built in the USA as described by Tuthill,[33] Vasta,[34] and Haviland.[27] The UK contributed a few coasters,[35] hundreds of barges,[36] a floating dock still in operation in Norway, and the unique Mulberry harbour pontoon complex used in the Normandy landings in 1944 and described by Wood.[37,38] The German shipbuilding programme has been reported by Finsterwalder[26] and Meyer.[39] Some barges and floating docks were built in the USSR during the war and the Soviets have continued to develop reinforced concrete for a wide variety of simple pontoon-type structures and barges.

Generally reinforced concrete, as distinct from ferrocement, has not been used for large vessels since the war except for some barges built in

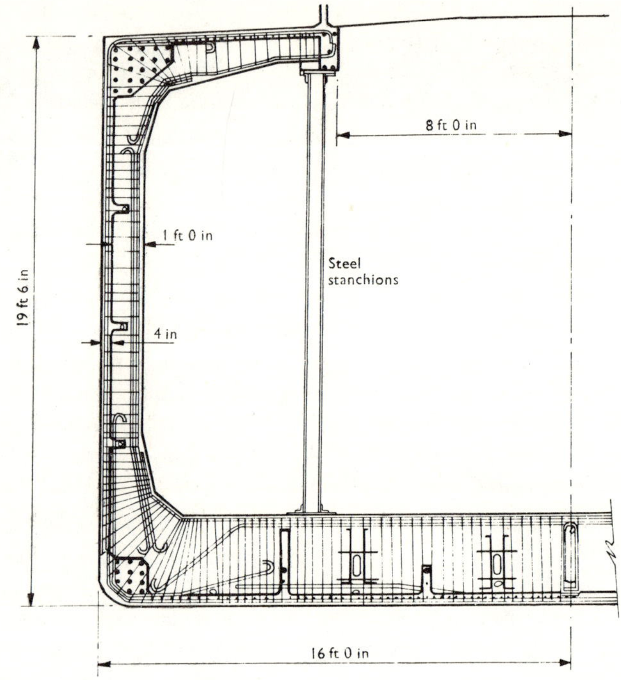

Fig. 6. Cross section of Armistice

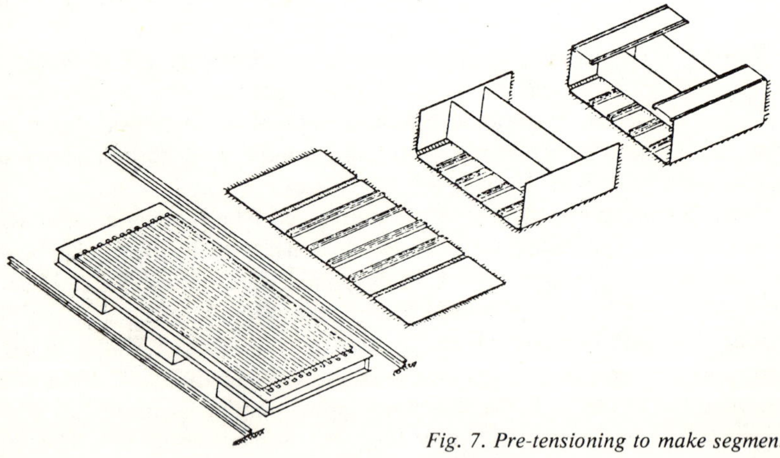

Fig. 7. Pre-tensioning to make segments

New Orleans,[14,40] pontoons and floating docks. For vessels over about 2000 t dead weight, prestressed concrete becomes more favoured.

PRESTRESSED CONCRETE

In prestressed concrete, precompression of the concrete is artificially and permanently induced to the same order of magnitude as the tensile stresses expected to occur under load. Precompression is applied by highly stressed tendons (high tensile steel wires, bars or cables) anchored to the concrete, and there are two basic methods of doing this.

In pre-tensioning methods the tendons are stressed and anchored, the plastic concrete being compacted around the tendons. When the concrete has reached sufficient strength the tendons are released from the anchorages, and as they attempt to regain their original length a compressive stress is induced by the bond or adhesion between tendons and concrete.

Alternatively, the concrete is cast first and the tendons, arranged either outside the concrete or more usually lying in preformed ducts within the structure, are tensioned when the concrete is strong enough and anchored at their extremities. The ducts are usually grouted with a cement slurry to protect the tendons. This process is known as post-tensioning and is more versatile because a series of prefabricated segments can be stressed together to make a complete structure.

Balancing of the prestress and the stresses caused by applied loads is most economically achieved by post-tensioning methods, as tendons can be curved and introduced into the structure where needed, thus varying the position and magnitude of the prestressing force.

Prestressing was first applied to vessels in 1943 when German engineers designed and built some 500 t barges.[41] Similar methods are adopted[42] by the Russians. Fig. 7 shows flat pre-tensioned slabs folded

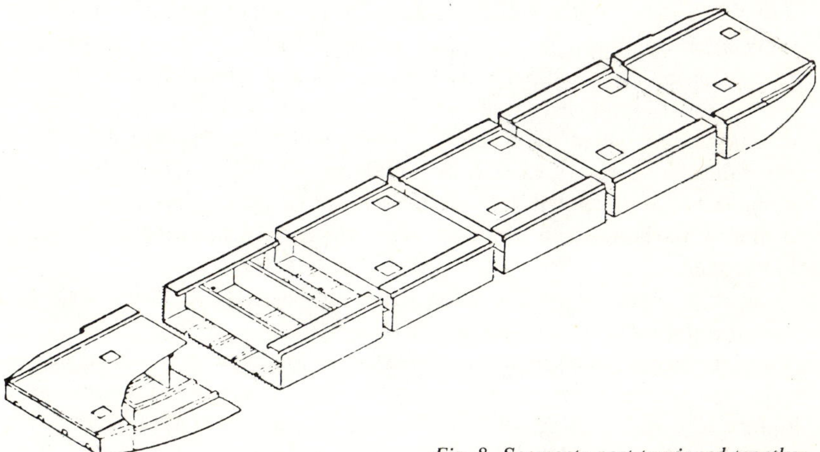

Fig. 8. Segments post-tensioned together

up to make simple box sections, which are then post-tensioned to complete the barge (Fig. 8). Future vessels are likely to be prestressed longitudinally, transversely and vertically.

Prestressing demands high quality materials and workmanship but it also means thinner and therefore lighter sections can be used. A pontoon of cellular construction was built in le Havre[43] in 1951; a 55 m × 24 m post-tensioned pontoon was built in Belgium in 1962 and towed to Libya as a floating oil refinery;[44] and 2000 t Philippine barges were built by Al Yee[45] in 1962. Many pontoons, barges and other craft have been successfully built in the USSR,[46,47] Australia, New Zealand and the UK.

EXPERIENCE OF SHIPBUILDING CONCRETE

Concrete lacks the wide range of long term operational experience but several important features in comparison with steel need to be examined. These are based on the experience of previous shipbuilding practice, contemporary reports from sailors and more recent research and development studies. Many of the features are similar for ferrocement and reinforced and prestressed concrete.

Durability

The durability of any type of reinforced concrete has to be considered under two headings—possible chemical deterioration, externally due to the aggressive sea water environment and internally because of reaction between dissimilar metals, and physical damage as a result of impact, abrasion, fire, explosion and freeze–thaw conditions.

In sea water, concrete is subject to sulphate and chloride action, but the reinforcement is usually adequately protected by even a thin cover of the highly alkaline cement mortar or concrete; this is borne out by the survival of many old vessels. Lambot's ferrocement boat of 1849 is still in excellent condition, Boon's *Zeemeeuw* of 1887 was in use until 1968, the reinforced concrete barge *Creteravine* is still afloat after 58 years of service and *Selma*, also built in 1919, demonstrated the high quality of her expanded shale lightweight aggregate concrete hull when test specimens were removed and examined in 1953.[48] Anyone visiting the Normandy beaches will see how well the concrete pontoons of the Mulberry harbour have survived over 30 years in the severe spray and tidal zone.

Floating docks show the most dramatic savings, requiring less than 10% the maintenance of similar all-steel docks.

To achieve this degree of durability the concrete must be rich in cement and low in water, with aggregates small enough to compact fully. Failure to observe these rules resulted in the poor condition of *Armistice* when surveyed in 1969, 50 years after launching.[31]

More recent experience with ferrocement vessels has shown that galvanitic action between galvanized and black steel reduces the strength of a hull, and small quantities of chromium trioxide (CrO_3) are needed to prevent this.[49] Evidence shows that any hull fittings must be completely isolated from the reinforcement, and over-protection by sacrificial anodes can result in deposits forming on the reinforcement, causing expansion and spalling of the surrounding concrete. The *Bulletin of the New Zealand Ferrocement Marine Association*[50] gives several case histories of corrosion problems and preventative measures.

Many vessels have been subjected to freeze–thaw cycles, and no deterioration is likely to occur where high quality concrete is used.

Damage and repair

Abrasion could be a problem, but a report[51] on a US ferrocement landing craft built in 1943, after many hard beachings under simulated wartime conditions said: 'after one and a half months of the toughest kind of tests . . . the concrete was unscathed—the original form marks were still visible'. Many ferrocement hulls seem to have avoided ice damage and Fougner was particularly pleased with his 1917 tug *Staal-beton 19* and says 'that after three or four winters in service, partly as an ice-breaker, there are no signs whatever of abrasion or wear on the concrete bows. The hull is in perfect condition everywhere.' However, some ferrocement lobster boats examined by the author[14] a few years ago had badly abraded bows and inner gunwales. If concrete is to be employed for boats working in very rugged conditions, far more attention in design must be paid to operational use.

The resistance to impact of ferrocement and reinforced and prestressed concrete needs much further study, for although most of the evidence suggests that those vessels which suffered major collisions or groundings did not come off too badly, the overdesign of such vessels must be taken into account. The author has seen several ferrocement vessels badly holed in the bows, but other reports are favourable. Dinsenbacher and Brauer,[8] writing about the Navy's CRAB (cement river assault boat) in 1973, said: 'Damage to ferrocement rarely results in breakage of the reinforcement, even when large areas of mortar are crushed. Severe impact may crack a large region, but the reinforcement usually retains much of the broken mortar, so that water leakage is less than would be the case for other materials.'

Fougner gives detailed reports on damage and repair of concrete ships and Harper[52] of the US Shipping Board gives an account of *Latham* (6380 t) after she had driven on some rocks at 10 knots and gouged out concrete in four strips running nearly the full length of the ship. She was dry-docked, concrete was removed, bars were straightened, fractured edges of the concrete were cleaned and grouted and fresh concrete was rammed and agitated inside timber forms—no leakage or even moisture was noted on refloating.

Elmont[53] recounts a similar story for the Danish vessel *Bartels* (1800 t) which had a comparatively thin 90 mm outer skin on a lattice girder double bottom. Repair costs were estimated at 50% of those for a steel ship exposed to the same impact damage.

However, impact is regarded as a major problem, although, unlike steel and timber, the material is unlikely to be split or holed by a severe localized impact load. The concrete usually shatters, with radial and circumferential cracking over a limited area, but the bent bars and meshes hold the pieces together so that leakage is manageable. Sometimes the external impacted surface may show only some slight indentation and apparently insignificant damage, whereas the inside surface is severely damaged with major spalling of the concrete. A further impact on the same area could completely disintegrate the hull, and it is therefore important that the insides of hulls should be easily accessible to inspection in those regions most likely to suffer collision.

The method for repair is simple, but it is essential that all fractured material is removed, the material is cut back to sound areas, the bars and meshes are cleaned and straightened and mortar (probably containing some epoxy bonding agent) is applied to make good.

Fire and explosion

All types of concrete are far less affected by fire and heat than unprotected steel, and wartime brought the additional hazards of bombs and mines. Meyer[39] reports that in 1944 a 1000 t German concrete barge hit a mine which exploded under the stern and the vessel was able to reach shore, being repaired while afloat with underwater concrete. An area of 12 m^2 of concrete was damaged but the reinforcement remained intact. A steel barge of the same size sailing alongside received an equal shock, sprung a leak, broke apart and sank. Other vessels also suffered from bomb and mine damage and were repaired quickly and effectively. Three bombs exploded within 10–15 m of a 300 t freighter—the concrete hull stood the strain undamaged but the engines were torn out of their foundations.

More dramatically, Turner[54] reports how a reinforced concrete barge survived the Bikini atomic test when its cargo of fuel oil was set alight.

Fire tests were carried out on an 8·2 m launch, 16 mm thick, when several gallons of petrol were allowed to burn for eight minutes in the sump and bilges. The hull was reported[55] as being in sound condition, its structural integrity maintained with no damage to the outside paintwork. There was some surface spalling of the mortar in the areas immediately adjacent to the fire.

Watertightness

There is no evidence of leakage in any operational hull, and reports emphasize that ships remained tight and dry even under the worst conditions. Tuthill[33] criticizes the classification rules requiring even small damp spots which appeared in hydrostatic tests to be repaired when it was known that such leaks would seal themselves in a very short period. These rules had been inherited from steel ships and unnecessarily added to delays in production and costs.

Vibration

Concrete is a good damping material and many sailors comment on the unusual pleasures of sailing in ships almost free from vibration.

Condensation

Weeping and condensation is virtually absent and bulk wheat stored within bare concrete hulls remained in perfect condition.[33] 'There is no condensation in the holds and as a consequence their cargo is kept particularly dry.'[56]

Seaworthiness

Doubts about seaworthiness have been expressed only by those who have never sailed in a concrete vessel. Seaworthiness was demonstrated by the splendid performance of the *SS Aspdin* as it headed into the hurricane off Cape Hatteras, in 1944. According to the captain's report,[57] the ship behaved admirably through wind of estimated speed 193 km/hour and waves of height 15·25–30·5 m. She pitched somewhat, but there was no rolling, panting, weaving or pounding. Such comments are repeated time and time again.

Design and strength

The strength of reinforced concrete hulls has never been questioned and there is no reason to doubt that adequate strength and safety can be achieved using modern design and construction methods.

Almost all experience with concrete boats and ships has involved copying the design shape and structure of timber or steel vessels, and the temptation to compare every hull material with steel is a very real one.

Vessels in both World War periods were framed: Fig. 6 shows a typical transverse frame, and Fougner[1] and Vasta[34] give details of design procedures for reinforced concrete. The Germans experimented with both framed and shell methods of construction in 1944 for vessels up to 3700 t and Finsterwalder[26] states that the framed vessels were not as suitable to resist impact loading as the shell form, especially as the hull–rib connection creates shear problems. The shell must be kept free from bending stresses under impact by bracing so that the stresses are predominantly tension and compression. Although thicker shells were able to resist impact loads adequately, hull weight remained the major problem and high strength low density concrete is needed to aim at a weight comparable with that of a steel hull.

Hull weight

The overweight hulls of both World War periods has left a legacy of contempt for concrete as a contender to steel,[58] and the predictions of designers were unrealistic.

The facts contradict the claims that, because scantlings are rather large for concrete, a bigger vessel means a lighter hull in comparative terms. Experience shows that hulls have been consistently overweight and no improvement in hull weight was made between the wars despite 20 years of experience in the design and construction of land-based structures, higher permissible concrete and steel stresses, and better quality control, handling equipment and methods of compaction.

In defence of these vessels, they were built under emergency conditions, the designers were under pressure to produce safe, seaworthy vessels, little attempt was made to fully utilize the material, and caution produced exceptionally strong but heavy hulls. Thousands of tonnes of steel were saved for essential war products. The vessels did what they were designed to do—and that was not to project concrete as a commercial alternative to steel as a hull material.

Soviet studies[42] show that hulls may be reduced by 20–25% using prestressing with ordinary concrete and by 40–45% using lightweight aggregate fillers.

Ferrocement builders claim equivalent hull weights for vessels over about 12 m but large ferro hulls have yet to prove themselves in operation over a long period.

High strength, lightweight aggregate concretes, reducing densities by only 16 kg/m^3, would have made a saving on hull weight of 400 t for the 5200 t (dwt) vessels which contained about 2300 m^3 of concrete. Concrete for *David O Saylor* would have needed to be about 1700 kg/m^3 to meet the estimate for that vessel.

While it is possible to speculate on the effect of producing high strength low density concretes, the stringent demands for high quality control in vast quantities of concrete cast in thin sections suggest that, until vessels of the sizes envisaged are actually built and costed, only little reliability should be placed on estimates.

Hull weight is not necessarily the sole criterion for estimating a ship's cargo efficiency, but it is a major problem which must be tackled realistically if concrete is to compete successfully for ships' hulls.

Quantity of reinforcing steel

The expected savings on steel in the wartime shipbuilding programmes were never achieved and it would appear that, over about 2000 t, unstressed reinforcement ceases to be economical and prestressed concrete gradually takes over as the more favourable structural medium. This fact is more quantitatively expressed in studies carried out in the USSR[42] which show that the quantity of steel can be cut in half using prestressing techniques.

It was anticipated in World War I that a concrete vessel would use about 33% of the steel needed for an equivalent-sized all-steel vessel, whereas the actual figure was nearer 60%.

Initial construction costs

The main purpose of the emergency shipbuilding programmes was to save plate steel, and while attempts were made to build economically the estimated costs were always far exceeded. Rising costs per deadweight tonnage with increasing size of vessel were very apparent, but in both war periods the concrete shipbuilding programmes were started towards the end of hostilities and the programmes halted before the expertise gained in construction could be fully utilized.

Costs are difficult to assess in comparison with those for all-steel vessels, but the estimates for the World War II vessels according to Tuthill[33] were $200 per tonne dwt and the actual cost $280. The repetitive use of moulds and accompanying experience did, however, produce significant reductions from $300 to $135 per tonne dwt during the production of 22 similar-sized (6375 t) oil tankers, while the most costly vessels, the 24 self-propelled tankers (5200 t) built at Tampa, fell from $410 to $314 per deadweight tonne.

Barges built at Bahia in 1911 cost $719 as against quoted steel barges at $1782. Barges in Panama in the same period were built at half the cost of steel ones and modern Soviet floating docks were 60–70% the cost of all-steel construction.

A Soviet report[42] gives the cost of construction of a 600 t barge 12% lower than steel in prestressed concrete and 3·5% lower in reinforced concrete.

In ferrocement construction the 'home built' boat is cheap in materials but labour costs would be high if these were evaluated realistically. Ferrocement is still a highly labour-intensive technique in laying-up of meshes and in many cases hand plastering. Developments of guniting or casting between strong moulds reduce labour costs and the potential of steel fibre reinforced mortars laid over a strong framework has yet to be fully exploited.

Costs depend much on local conditions; savings can certainly be made in developing countries where there is a vast pool of unskilled labour, expensive dockyard equipment associated with steel shipbuilding is not necessary, smaller quantities of steel need to be imported and the basic materials for cement and aggregates are readily available all over the world. Coupled with this are the social and economic advantages of providing employment.

It will be more difficult to reduce initial building costs in highly industrialized, steel-orientated countries and it must be remembered that the hull is only part of the total cost of a completed vessel.

Certainly prefabrication and mass production techniques will be developed for standard vessels and this could bring real economies in construction costs.

Operational costs

Heavier hulls and possible increased skin friction[59] will make operational costs higher, although reports suggest that better steerage and seaworthiness in heavy weather compensate for this. Marine growth could be less on a concrete hull as toxic materials can be introduced into the concrete.[60]

Any costing of concrete hulls must include the benefits of low maintenance and easy repairability, generally without the need for expensive and time-consuming docking.

Hull shape

A concrete boat built in 1887 adopted the clinker lines of a timber vessel, and this complex of concrete not identifying itself has been repeated time and time again. Steel hull shapes and structural framings have been copied and duplicated and rarely has any attempt been made to take full advantage of the constructional versatility of concrete; details of various shapes are given elsewhere.[58]

Concrete needs to be moulded in simple standard shapes for segmental construction and slipforming techniques to make as much use of repetition as possible, and this will not necessarily produce the most efficient shape for low resistance through water, so an economical compromise must be made. The hard commercial reality of ship operations means that the concrete must be at least half the cost of steel to make it a viable proposition.

The concrete ship designer must evolve shapes for vessels based on sound knowledge of the behaviour of concrete under the complex loading conditions imposed on a floating hull, the best use of the variety of construction techniques possible and a realistic assessment of operational and other costs. For too long the concrete vessel has relied on the traditional steel concept: concrete ship designers must start from fundamentals if concrete is to become recognized as a structural medium for ships' hulls in its own right.

Construction

All the studies made by the author of past vessels show that almost every conceivable method of structural shape and building technique has been tried but with little success.

Double-walled skins were first used in 1918 and offer strength and lightness; slipforming was tried in 1915 and prefabrication widely adopted in World War I.

Vasta's[34] description of an in situ tanker requiring 2300 m^3 of concrete, taking 500 000 separate pieces of bar reinforcement amounting to 6·5–8% by volume, 475–590 kg per cubic metre of concrete with 23 200 m^2 of formwork shows the enormity of constructing a relatively small vessel of some 5200 t.

Prestressing points to an economical and durable material which could be successfully used in medium-sized ocean vessels where post-tensioning provides longitudinal continuity and strength, while the highly resistant ferrocement skin gives adequate local strength. As yet there is little evidence of the success of the long term operational qualities of ferrocement, whether used alone or in conjunction with prestressing.

Simple constructional techniques are essential, possibly using peripheral prestressing on double-walled ferrocement or fibre-reinforced concrete walls for transverse and local strength, fore and aft post-tensioning providing the longitudinal strength.

Classification societies

Government agencies and the classification societies had every justification for being cautious when asked to adopt reinforced concrete for ships' hulls. Fougner relates how he was expected to design his ships 20% stronger than equivalent-sized steel vessels, and L. Coff, a consulting engineer from New York, was very adamant when writing in 1944[61]: 'No progress in concrete shipbuilding can be made if the boats have to be classified in the normal way. It is well nigh impossible to do an efficient pioneering job and to have the classification company, going only by experience, endorse the ingenuity of the designer (if any) by a certificate. It is unfair to expect them to.' And again in 1945[62]: 'The worst thing was to have to fight the mental attitude of the classification companies. In the former emergency their surveyors, distinguished and experienced naval architects, bought their first text books on reinforced concrete the very day they started putting down the law and telling the concrete engineer what to do and what to leave alone. Reinforced concrete was then less than 20 years old as an industrial proposition . . . No concrete boats had been built since 1922. What more natural than that classification companies should insist on starting where we left off then?'

Tuthill[33] concedes these points: 'It is true and admittedly regrettable that the four designs for the 77 ships built in five yards did not differ very much from the designs for the concrete ships built at the time of World War One. As Mr Coff has correctly surmised, the classifying agency had much to do with this due to its refusal at the start of the program to permit any departure from previous standards and concepts of design because there was no time to prove the many new factors involved ... Consequently the concrete ship program was automatically burdened with classification and the many resulting needless and impractical requirements that well could have been ignored in producing safe, practical hulls of high utility in the emergency.'

Happily, classification societies and concrete ship designers and builders are co-operating in developing guide-lines and rules for ferrocement and reinforced and prestressed concrete, and each is learning from the other.

CONCLUSIONS

The present shipbuilding phase, like the first, is based on engineering and economic considerations and generally these are well justified both in construction and operation. The only experience with large vessels comes from the two war periods and those hulls were grossly overdesigned and overweight, used far more steel than was anticipated and never came up to the optimistic predictions in construction time and costs.

The reported sound behaviour of ships under load and their excellent seaworthiness must be considered alongside their gross overdesign, and more realistic calculations and structural sizes would produce more obvious deflections under load.

Experience of concrete hulls indicates two major areas for closer examination and caution in design—impact behaviour and weight. On the whole, impact resistance can be improved at extra cost; but, unless the naval architect moves to shapes more able to resist the complex stresses at sea and uses higher strength and lower density concrete, commercial concrete ships will not be able to compete operationally with steel vessels. Concrete engineers and naval architects must grasp this nettle firmly and be realistic in their predictions of initial cost, construction time and self weight because these are three important considerations in the economic evaluation of a concrete vessel.

At one time calculations indicated a break-even point in self weight with steel ships at about 100 000 t, but the author now believes that seagoing concrete ships will always be 10–15% heavier than steel unless some major, and inexpensive, method is found to produce very high strength, low density concrete. Equally, where the hull constitutes the major financial part of the complete vessel as in large barges, concrete construction should aim at 70% the price of all-steel hulls. Where mass prefabrication techniques can be used in a fleet of identical vessels, this figure could be reduced by 60%.

The old prejudices about concrete ships had been overcome by the success of the amateur yachtsman and a few other bigger vessels long before the North Sea demanded concrete production platforms. The concrete ship designer needs to evaluate the total cost of building and operating a ship over a period of years to compare with steel ships for potential operators to assess.

The author is fortunate in serving on two committees studying concrete ships, the Concrete Society/Royal Institution of Naval Architects working party on ferrocement vessels and the Fédération Internationale de la Précontrainte Commission on Concrete Ships.[63] In these two small groups of people there are naval architects, classification societies and concrete specialists, and a refreshing and enthusiastic desire to produce safe and economical concrete vessels.

People and experience are important ingredients, and examination of the experiences of previous concrete shipbuilders, designers, operators and sailors and the knowledge of the sort of people involved in projecting concrete ships today should produce a confidence and prospect for the potential of concrete hulls which has not yet been realized.

REFERENCES

1. FOUGNER N. K. *Seagoing and other concrete ships*. Henry Frowde and Hodder and Stoughton, London, 1922.
2. NEVILLE A. M. *Properties of concrete*. Pitman Publishing Ltd, London, 1975. 2nd revised (metric) edition.
3. CASSIE W. F. Lambot's boats. *Concrete*, 1967, **1**, Nov., No. 11, 380–381.
4. Lambot's boat. *Concrete*, 1977, **11**, Feb., No. 2, 24–25.
5. BATTANDIER A. Une barque en ciment. *Cosmos, Torino*, 1897, **36**, 718.
6. BOON A. A. *Der ban von schiffen aus eisenbeton*. W. Ernst, Berlin, 1917.
7. NERVI P. L. Ferrocement: its characteristics and potentialities. *Ingegnere*, ANIAI, Italy, 1951, **I.**
8. DINSENBACHER A. L. and BRAUER F. E. Material development, design, construction and evaluation of a ferrocement planing boat. *Marine Technology*, 1974, July, 277–296.
9. *Conference on fishing vessel construction materials, Montreal 1968*. Dept. of Fisheries and Forestry, Ottawa, 1969, Canadian Fisheries Report No. 12.
10. *Seminar on the design and construction of ferrocement fishing vessels, Wellington, New Zealand, 1972*. United Nations, Rome, 1973. FAO Fisheries Report No. 131, Vol. 11.
11. *Concrete sea structures*. Fédération Internationale de la Précontrainte, Wexham Springs, Slough, 1973. Proceedings of symposium held at Tbilisi, USSR, 1972.
12. CANADIAN INDUSTRIAL DEVELOPMENT BRANCH. *Ferrocement for Canadian fishing vessels*. Dept. of Environment, Ottawa. Project reports: No. 42, Aug. 1971; No. 48, Jan. 1972; No. 55, Aug. 1972; No. 64, June 1973.
13. GREENIUS A. W. *Ferrocement for Canadian fishing vessels—a summary and interpretation of test results—1969–1974*. Industrial Development Branch, Fisheries and Marine Service, Environment Canada, Ottawa, 1975. Technical report 86.
14. MORGAN R. G. *Concrete floating and submerged structures*. The Concrete Society, London, 1973.
15. MORGAN R. G. *Ferrocement—a summary of research activity*. National Physical Laboratory, London, 1974, 111–112.
16. BOWEN G. L. and BAUGH L. C. Ferrocement construction: topics for research. *Naval Architect*, 1975, Apr., 53–56.
17. FYSON J. *Construction of a 16 m ferrocement fishing boat*. FAO, United Nations, Rome, 1970. FAO Fisheries Technical Paper No. 95.

18. BIGG G. N. *An introduction to design for ferrocement vessels.* Industrial Development Branch, Canadian Dept. of Environment, Ottawa, 1972. Project report 52.
19. JOHANSEN J. K. *Strength calculations of ferrocement for ship hull constructions.* Det Norske Veritas, Oslo, 1973. Report no. 73–84 (in Norwegian).
20. LACHANCE L. and FUGERE P. Construction of a ferroshotcrete motor sailer hull. *Ingénieur, Montréal,* 1970. Mar.
21. BEZUKLADOV V. F. *et al. Ship hulls made of ferrocement (design, strength and construction technology).* Shipbuilding Publishing House, Leningrad 1968. (Nav-ships translation No. 1148, Dept of Navy, Washington.)
22. SEARLE A. B. Reinforced concrete ships, barges and pontoons. *Concr. constr. Engng,* Supplement, 1918, Nov. and Dec.
23. HØJGAARD K. The development of reinforced concrete shipbuilding during the war. Paper read to Danish Society of Engineers, Feb. 1920.
24. MORGAN R. G. *Concrete,* 1972, **6,** Sept., No. 9, 28.
25. JONES L. and TURNER F. H. Britain's first concrete ship? *Concrete,* 1972, **6,** July No. 7, 26–27.
26. FINSTERWALDER U. Schalenschiffbauten. Proceedings of 45th congress of Deutscher Beton-Verein (EV), Wiesbaden, 1948, 96–112. Translated by J. Perry, Univ. Bristol.
27. HAVILAND J. American concrete steamers of the First and Second World Wars. *Am. Neptune,* 1962, **22,** No. 3, 157–183.
28. DENNY M. A preliminary survey of the possibilities of reinforced concrete as a material for ship construction. *Trans. Instn Nav. Archit.* 1918, **60,** 123–137.
29. POLLOCK W. Reinforced concrete vessels. *Trans. Instn Nav. Archit.,* 1918, **60,** 138–153.
30. THURSTON T. G. O. On the design and construction of a self propelled reinforced concrete seagoing cargo steamer, now building in Great Britain. *Trans. Instn Nav. Archit.,* 1918, **60,** 154–187.
31. MORGAN R. G. Obituary of a pioneer: the coaster SS Armistice. *Concrete,* 1970, **4,** Feb., No. 2, 59–61.
32. Reinforced concrete shipbuilding in Dorsetshire. *Engineer,* 1918, **126,** 15 Nov., 408–416.
33. TUTHILL L. H. Concrete operations in the concrete ship program. *Ind. Am. Conc. Inst.,* 1945, **16,** 137–177.
34. VASTA J. The concrete ship program of World War II. Paper to Chesapeake Section of American Society of Naval Architects and Marine Engineers, 1952.
35. MORGAN R. G. (Brooks A. E. (ed.)). Concrete for shipbuilding: a challenge to lightweight concrete. *Proceedings first international congress on lightweight concrete.* Cement and Concrete Association, Wexham Springs, Slough, 1968, **2,** 180–184.
36. Production of concrete barges. *Engineer,* 1944, **177,** Mar., No. 4603. 253–254.
37. WOOD C. R. J. Phoenix. *The civil engineer in war.* Insitution of Civil Engineers, London, 1948, **2,** 336–369.
38. WOOD C. R. J. Reinforced concrete pier pontoons and intermediate pierhead pontoons. *The civil engineer in war.* Institution of Civil Engineers, London, 1948, **2,** 401–450.
39. MEYER G. A. *German concrete shipbuilding during the war.* His Majesty's Stationery Office, London, 1946. FIAT final report No. 844.
40. *Concrete barges multiply in Gulf.* Concrete Products, Chicago, 1967.
41. WALLEY F. *The use of prestressed concrete in Germany.* His Majesty's Stationery Office, London, 1946. B105 final report No. 1712, 17.
42. KUDRYAUTSEV A. A. The use of prestressed concrete in shipbuilding. *Sudostroenie,* 1960, **26,** 44–47. (National Lending Library, Boston Spa, NLL RTS 5392)
43. Reconstruction of quays at Le Havre. *Concr. Constr. Engng,* 1951, **46**, Apr., 111.
44. SMET V. and CAMPAS P. A floating refinery built on a pre-stressed concrete barge. *Bull. Tech. Bur. Veritas,* 1973, **45,** 158–167.
45. YEE A. L. *et al.* Design and construction of oceangoing pretensioned concrete barges. *J. Am. Concr. Inst.,* 1975, Apr. 125–134.
46. SEREBRENNIKOV B. N. Pontoon barge of prestressed dense concrete. *Sudostroenie,*

1963, **29,** 43–44. (National Lending Library, Boston Spa, NLL TS 5407)

47. EGOROV N. M. Concrete applications in vessels for inland navigation. *Concrete sea structures*. Fédération Internationale de la Précontrainte, Wexham Springs, Slough, 1973. 233–235.
48. *A report of an investigation on the condition and physical properties of expanded shale reinforced concrete after 34 years exposure to sea water*. Expanded Shale, Clay and Slate Institute, Washington DC, 1960, 2nd ed.
49. CORNER I. *et al.* Chromate admixture to improve performance of galvanised steel in concrete sea structures. *Concrete sea structures*. Fédération Internationale de la Précontrainte, Wexham Springs, Slough, 1973. 159–163.
50. *Journal of Ferrocement*. New Zealand Ferrocement Marine Association, PO Box 26–073, Auckland 3, NZ.
51. MACLEAY F. R. Thin wall concrete ship construction. *J. Am. Concr. Inst.*, 1949, **21,** Nov. No. 3, 193–204.
52. HARPER W. P. The successful repair of a concrete ship bottom. *Concr. Constr. Engng*, 1921, **16,** Dec., No. 12, 811–812.
53. ELMONT V. Repairing a concrete vessel. *Concr. Constr. Engng*, 1921, **16,** Sept., No. 9, 597–603.
54. TURNER F. H. Initial report on the resistance of concrete hulls to fire and explosion. Submitted to meeting of Fédération Internationale de la Précontrainte Commission on Concrete Ships at Berkeley, California, 1975.
55. SUTHERLAND W. M. *Boats from ferrocement*. United Nations, New York, 1972
56. LUZON STEVEDORING CORPORATION. Prestressed concrete barges v. steel barges: report of the operators on maintenance and repair costs. LSC, Republic of Philippines, 1970. Internal report.
57. ROGAN E. Description of experiences during the hurricane of September 14, 1944 as encountered by the *SS Joseph Aspdin*. Private report by the ship's captain, 1944.
58. MORGAN R. G. History of and experience with concrete ships. *Proceedings of the conference on concrete ships and floating structures*. University Extension, University of California, Berkeley, 1976, 3–16.
59. MIKHAILOV V. N. and VULODIN S. A. Determining the frictional resistance of a reinforced concrete ship. *Sudostroenie*, 1961, No. 2, 5–10. (NLL RTS 5408).
60. KEEN R. *Controlling algae and other growths on concrete*. Cement and Concrete Association, Wexham Springs, Slough, 1971, Advisory note 45.020.
61. COFF L. Design of concrete ships. *Engng News Rec.*, 1944, **133,** Dec., 76.
62. COFF L. and TUTHILL L. H. Discussion on Concrete operations in the concrete ship program. *Ind. Am. Conc. Inst.*, 1945, **16,** Suppl., 180–183.
63. FÉDÉRATION INTERNATIONALE DE LA PRÉCONTRAINTE COMMISSION ON CONCRETE SHIPS. *Prestressed concrete ships first report—considerations for design*. FIP, Wexham Springs, Slough, 1975.

Concrete afloat. Thomas Telford Ltd, London, 1977, 23–41

The durability and economy gained from the use of concrete for ocean-going vessels, coupled with recent experience of gravity structures for the North Sea oil fields involving long deep-sea tows of structures in excess of 250 000 t, has brought interest to the construction of complete production plants erected inshore on floating platforms for trans-shipment complete to their destinations. Considerable advantages arise from this technique in overall cost, shorter and more positive construction schedules and large reductions in the demand for expatriate skilled labour. For the structures, prestressed concrete designed for the marine environment is the natural construction material, for reasons of economy, durability and ready availability. The paper discusses the potential for this type of structure, the design methods recommended, construction procedures, and towing and installation methods for typical structures.

Prestressed concrete platforms for process plants

J. A. DERRINGTON, BSc(Eng), FICE*

The large increase in demand for oil-related industrial developments and associated works in the developing countries has placed considerable strains on their local construction industries. The trans-shipment of the materials required and the engineering equipment which is to be installed has often overloaded the local port handling facilities, leading to delays in construction schedules. The demand for highly skilled labour in areas where this is not normally available has required large amounts of expatriate labour and the establishment of accommodation and support facilities for this labour. Not surprisingly, this has led to considerable increases in unit costs, compared with those of the developed countries in Western Europe and North America, labour rates often increasing threefold.

At the same time, the developing technology in the North Sea and other areas of offshore working has shown that massive concrete structures with total operational weights of 300 000–400 000 t can be constructed partly onshore, floated out and completed close inshore in sheltered deep water. Their topside engineering facilities can then be installed in a single operation under sheltered conditions, before tow and installation to hostile marine environments, without difficulty or delay. The replanning of the engineering plant within limited deck areas (25 000 t within a deck area less than 100 m square) has been found to be a natural development in process plant technology. Installed platforms have been constructed to support the development drilling of a 100 000 barrel/day oil reservoir, the primary separation of the product, pumping ashore, the

* Director and Head of Design Group, Sir Robert McAlpine & Sons Ltd, London.

generation of sufficient power to support the facility, the processing and injection of water to maintain reservoir pressures or the reinjection of the separated gas, together with the provision of accommodation for a labour force in excess of 100 men. All these operations, which a few years ago lacked credibility to a potential client, have now been proven,

Fig. 1. Floating oil refinery, Refiner I

Fig. 2. The ARCO LPG terminal

and suitable ocean-going tugs for long voyages and the necessary insurance cover have become readily available.

It is not surprising that plans have been prepared for process plants of many kinds, using the technology of complete prefabrication and commissioning, in localities where engineering and labour resources are available. Already such units have been installed on the coasts of North Africa, the Arabian Gulf and Indonesia. The natural material for the structure is prestressed concrete, offering over other materials the advantages of lower initial cost, greater durability and fire resistance and the ability to operate either in a floating or grounded mode without the need for additional foundations or expensive maintenance work. The ability to remove the total structure, should basic raw material patterns change, is in the eyes of the environmentalist an added attraction.

APPLICATION

Many applications of this technique on steel or concrete structures already exist. An early one was the construction of a small petroleum refinery for Société Esso-Standard at Tamise, Belgium, and its installation at Marsa el Brega, Libya, in 1962 (Fig. 1).[1,2] Of more recent interest was the construction in Tacoma, Washington, USA, of a 60 000 t floating barge for Atlantic Richfield for the liquefaction and export of liquefied petroleum gas (LPG) and its installation in the Ardjuna field, Indonesia, in 1976 (Fig. 2). However, many more applications have been reported recently, including a 1500 t/day ammonia plant and a 1700 t/day urea plant for Pertamina, a 250 000 t/year paper pulp plant destined for Brazil and numerous applications for liquefied natural gas (LNG) liquefaction and storage.[3]

Process plants for the petrochemical industry are not the only application to have received attention. Studies have been completed for plants for desalination, water treatment and injection, electrical generation from gas turbines, offshore cold storage, mechanical workshops and water treatment/sewage digestion/electricity generation for new township developments and large hotels with completely integrated mechanical services.

Compared with an oil production platform measuring 100 m square in plan, weighing 350 000 t and installed in 150 m depth of water, many of these structures appear small. For instance a desalination plant with a 9000 t/day capacity can be constructed upon a concrete platform 75 m long by 40 m wide and 7·5 m deep. This will have a loaded draught of less than 5 m; and including the electrical generating equipment and a boiler house, so that it can operate as a completely independent unit requiring only a fuel supply, it will weigh only 15 000 t and provide storage for 6000 m^3 of fresh water, or 6000 m^3 of fuel. The tow of such units (Figs 3 and 4) from Western Europe to the Arabian Gulf is a task

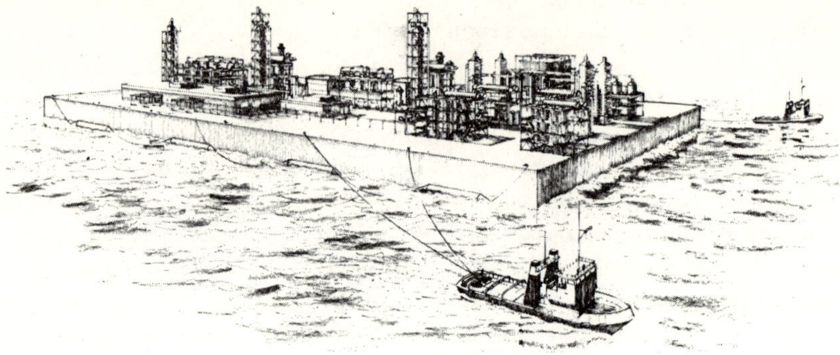

Fig. 3. Artist's impression of chemical plant under tow

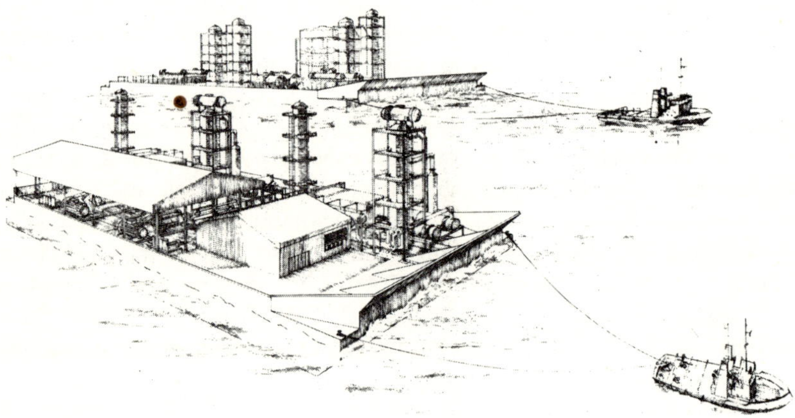

Fig. 4. Artist's impression of ammonia plant under tow

well within present experience, and most types of platforms of this nature can safely pass through the Suez Canal and Red Sea.

The concrete structure for such an application is divided by five longitudinal bulkheads and four transverse bulkheads into 30 compartments and requires a 45 cm thick base, a 25 cm thick roof and 45 cm thick exterior walls. Internal compartmenting walls, designed to withstand full hydrostatic pressure in case of a catastrophic collision, are each 30 cm thick. The advantage of construction in an area where high standards of workmanship and materials are readily available gives the further advantages for this application that the evaporator tunnels and storage areas can be also constructed of high grade prestressed concrete in place of steel and phosphor bronze, offering increased durability and lower material costs.

CONCRETE QUALITY

The quality of concrete used for structures permanently exposed to a marine environment is critical, for with normal structural grades the possibility of moisture permeating through the cover to the reinforcing bars or through cracks in an unstressed structure can lead to corrosion of the reinforcement. Even structures which after their installation are beached or installed in a permanent basin remain throughout their lives in an environment hostile to the reinforcing bars or post-tensioning tendons. The integrity of the concrete cover is therefore of great importance.

Although considerable experience exists of marine structures and concrete ships with great durability even though constructed without the benefits of modern concrete technology (e.g., Nab Tower,[4] Mulberry caissons,[5] coastal vessels[6] and many others), concern for the durability has led to increased thicknesses of cover being specified in the rules of the various ship classifying societies and national standards. The proposals of the Fédération Internationale de la Précontrainte (FIP)[7] suggest 75 mm and 100 mm for passive reinforcement and post-stressing tendons respectively in sections over 50 cm thick in zones suffering the greatest exposure, whilst ship classification society rules such as those of Det norske Veritas (DNV)[8] propose comparable figures. Recent experience has, however, highlighted problems arising from these large amounts of cover. Provided that assurance can be obtained of the quality of workmanship particularly in the concrete cover it is likely that these requirements will be reduced considerably. Experience in the use of fibre reinforcement in the cover zones to increase locally the toughness of the cover will be most helpful in reducing cover to the practice used in the early years of the twentieth century.

The use of special cements in the concrete to resist the attack of chemicals present in sea water has also attracted interest, but the main requirement for a densely compacted concrete containing good quality aggregates, a minimum cement content of 350 kg/m^3 and a water/cement ratio not exceeding the range 0·40–0·45 has been proved to be suitable. Certainly the use of special sulphate resisting cements has been found unnecessary and recent work has shown that the tricalcium aluminate (C_3A) content of the cement has little relevance to the durability of high strength concrete. Ordinary Portland cement of British manufacture, having C_3A content in the 10–12% range, has been proved suitable. The partial replacement of cement by fly-ash or other pozzolanic materials has been found satisfactory, although the reduction of the setting heat of the concrete by this means has little relevance in the structural sections normally chosen.

Lightweight aggregate concrete has obvious attractions for the larger structure due to the considerable saving in structural dead weight achieved (15–20%). For such schemes as a bulk LNG carrier, the increased thermal performance and lower elastic modulus can also offer advantages, and above all the proven record of lightweight concrete in ship construction is undisputed. However, the thinner sections possible with high strength concretes can reduce some of the apparent savings in weight, for lightweight concrete of suitable strength can seldom be produced with specific gravity less than 1·9. Above all, the need for proving its durability in cases where classification is necessary provides further work for the concrete technologist.

Other materials problems require consideration, not least the adequate protection of all exposed steelwork and the proper grouting and sealing of all post-tensioning tendons and anchorages. The high level of prestress required to resist the fluctuating stresses at sea requires concentrations of large tendons, and the anchorage zones for these require careful attention. The low level of stress variation in the tendons arising from frequent reversals of loads when afloat is one great advantage of prestressed concrete, and fatigue considerations rarely give rise to problems.

DESIGN BASIS

The development of limit state design philosophy as used in CP 110[9] has provided a useful method of relating the safety of a structure to the level of risk in the different stages through which it passes to completion:

(*a*) whilst grounded in the construction basin;

(*b*) when completed and afloat in sheltered waters but not yet operational;

(*c*) during its relatively short passage over the high seas to its destination;

(*d*) in an operational mode when finally beached or permanently moored in sheltered waters.

For each of these conditions it is necessary to consider the combination of partial safety factors required for serviceability and safety. But it is general when the structure is on passage to its destination that the most critical circumstances will prevail and it is this condition which requires the most detailed examination. For this purpose it is necessary to make as realistic an assessment as possible of the platform behaviour during tow, an aspect of the design which is probably more the speciality of the naval architect than the civil engineer.

The action of waves on a floating concrete platform needs assessment to determine, firstly, that the platform has adequate structural strength and, secondly, what its behaviour will be under various sea states.

Until comparatively recently, sea states have not been amenable to analytical treatment and the only design methods available have been empirical methods which experience has shown provide acceptable

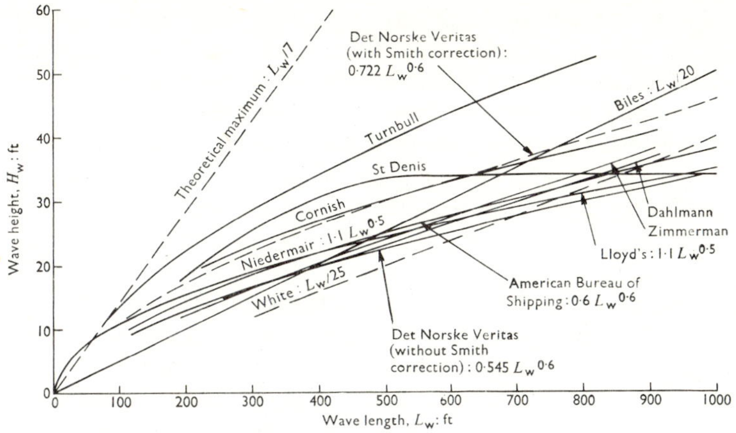

Fig. 5. Wave heights for ship bending moments[13, 14]

margins of safety for 'ship shaped' vessels. Generally, classification societies and regulatory bodies base their minimum strength requirements on their own rules, which vary. These are presented as formulae which take into account the vessel size and hull form and so provide guide-lines for calculating maximum values and distribution of bending moments and shear forces. There may also be stipulations governing minimum hull section modulus and maximum allowable length of structure. Wave-induced flexure and shear may also be estimated by considering the vessel as effectively balanced in hogging or sagging mode on a supposedly static 'design' wave of a length equal to the vessel length and of a height regarded by the authority as representing the most severe environmental condition to be encountered. A comparison of design wave heights from several authorities is shown in Fig. 5.

Rectangular platforms have a very different geometry from ships and to apply rules intended for ship forms is not always correct. However, the design wave principle is easily applied to a regular hull shape. Moments and shears are usually calculated assuming a symmetrical sinusoidal wave, and torsional moments assuming a quartering wave with successive wave crests at diagonally opposite corners of the platform. To these must be added the still-water moments and forces due to uneven distribution of load from payload or structure.

These methods, though valid, are conservative and more realistic results are obtained from statistical methods to which recourse must be made, in any event, to evaluate motion while under tow.

For design purposes the floating platform is regarded as being subjected to an irregular sea composed of broad-crested waves travelling in a direction parallel to either of the principal axes (pitch or roll). An irregular sea is considered to consist of a spectrum of randomly distributed regular wave components covering a wide band of frequencies.

Under storm conditions the greater part of the wave energy is concentrated in a relatively narrow band, the location of which is dependent upon wind intensity and storm duration. Mathematical models have been developed which approximate natural storm spectra and which enable a complete spectrum to be defined in terms of certain parameters, usually significant wave height and average apparent period. This is usually expressed in terms of energy (spectral density) as a function of frequency. It has the very useful property that, under certain conditions, it may be used in conjunction with an appropriate transfer function to provide a corresponding energy spectrum for any particular response. A required condition is that the amplitude of the response to a regular wave is linearly related to wave amplitude, in which case the response to irregular waves is obtained by effectively superposing the responses to individual regular waves. The method is then applicable to motions, moments, forces etc. The transfer function is obtained from an analysis of the response to a sufficient number of regular waves of different frequency within a frequency range matching that of the wave energy spectrum. Thus the computation has three stages:

(*a*) defining the wave spectrum;
(*b*) determining the response to regular waves;
(*c*) evaluating the response parameters.

A decision as to the critical wave spectrum must be based on a detailed study of meteorological and oceanographical data relating to the towing route, including seasonal variation and possible combinations of storm waves and swell. The duration of the tow must also be taken into account in assessing level of risk and probability of extreme values.

The response of a rectangular platform to regular waves, and hence the transfer function, may be obtained with sufficient accuracy using a simple two-dimensional strip theory. The procedure is to obtain an expression for the total vertical wave force acting on a cross-sectional element of the vessel. The components of this force are a hydrostatic force, an inertia force and the two components of hydrodynamic force involving added mass and damping. Wave forces are assumed to be undisturbed by the presence of the vessel. The force and moment due to the elements are then integrated over the length of the vessel and the response amplitudes are thereby evaluated.

Although most of the ship classification societies have already certified concrete barges as ships, some cases date back as far as World War I, and as far as it known there are no definitive 'rules' yet published dealing with prestressed concrete ships or platforms. Nippon Kaiji Kyokai have prepared a rule for prestressed concrete barges, an a provisional note by Bureau Veritas (BV) dated July 1976[10] based on the Society's rules for steel ships and the rules for offshore platforms, lists a number of requirements for prestressed concrete ships. It is also understood that Lloyds, DNV and American Bureau of Shipping (ABS) are working on draft rules covering this sort of structure. The main aspects of BV's document are summarized below insofar as they apply to process platforms:

(*a*) a reduction is allowed in the values of sea loads for objects destined to be towed only once; in this case the values are to be determined from a statistical calculation;
(*b*) loads are specified from the point of view of limit state calculations applied to concrete structures such as those defined jointly by the Comité Européen de Béton and FIP;
(*c*) several serviceability limit states are proposed; generally, these take the form of limiting crack widths according to the nature of any liquid containment requirement;
(*d*) two classes of environmental load are considered, normal and exceptional, corresponding respectively to return periods of 1 month and 50 years;
(*e*) maximum and minimum angle of roll and consequent forces are specified.

Currently, the imposition of an arbitrary limit to the length of concrete ships to reduce possible vibration effects is being discussed. The US Coastguard limit of 200 m is likely to be increased as there is some evidence that concrete structures are considerably stiffer than their equivalents in steel. This limitation would not appear to apply to concrete platforms, whether intended for beaching or for floating operation.

In addition to the design specification it is now customary to require the structure whilst floating to possess 'one compartment damage' stability to ensure that after a catastrophic collision or accident it will remain both buoyant and stable. Transverse and longitudinal bulkheads are designed at a suitable limit state for full hydrostatic pressure, and pipework, pumps and a standby source of power are available for restoring the floating trim of the structure in such an eventuality. This system, which must be installed before the structure is floated, must not only deal with any water shipped during the voyage, but also enable the final installation to be carried out under positive control if the structure is to be bedded down on its foundations at its destination.

CONSTRUCTION PROCEDURE

The construction of a concrete platform for an industrial plant differs from conventional ship construction chiefly in the size and shape of the structure, and this makes construction in a conventional dry dock virtually impossible on account of the width. Because of the effect on the concrete structure, construction on a slipway is difficult.

The experience gained in gravity platform production for the offshore oil industry is more relevant, and construction in a dry basin which can be flooded when at an appropriate stage is more common. The draught requirements for critical flotation are much less, generally not exceeding 5 m. Completion of construction whilst afloat is not always necessary as many locations around the UK coasts have suitable tidal ranges for floating out the completed structure into deeper water without restriction. The draught of the completed structure is seldom greater than 10 m and does not impose the problems of concrete oil platforms where 30–40 m deep sheltered water is almost a necessity.

Construction work can be carried out within dry basins formed on the foreshore and protected from the sea by steel sheet-pile cored earth walls. The base slab is constructed on a suitable permeable subsoil and the external walls and bulkheads are constructed of prestressed concrete using specially produced sectional forms. Pipework and other engineering services within the hull are installed after the post-tensioning of the concrete structure has been completed. The deck slab construction may be completed within the dry basin, which will then be flooded on a suitable tide by removing the temporary enclosure wall. Installation of the necessary mooring and towing points, and provision of the pumps, pipework and valves for controlling the trim of the structure when under tow and to provide safety and stability should accidental damage occur, will take place in the dry dock. The structure will then be towed to a nearby fitting-out wharf where the major part of the mechanical engineering works will be installed. The flat base slab and cellular construction has inherently considerable strength and stiffness, and allows the facility for beaching the completed structure alongside the fitting-out wharf and providing a stable foundation for the plant installation.

Fig. 6. Prefabricated crew accommodation module

Mechanical works are prefabricated whilst the hull construction is proceeding to optimize the construction schedule. This is already a natural development of the process plant industry. The ability to use large floating cranes or dockside cranes to install packages of 100 t weight or more has been found of considerable benefit to both programme times and economy and has, with North Sea platform experience, become accepted practice. Alternative methods of skidding onto the deck even larger pieces of equipment, assembled ashore, offers similar advantages.

The fitting out and plant assembly at a separate location from the construction basin offers some advantages, in that the co-ordination of the plant installation with the concrete works on one site could produce delays, and the different types of labour and union organization do not readily integrate on one site.

TOWING AND SEAFARING

Certification has been covered in the design section, for although not currently required by legislation, advantages obviously accrue from the normal certification process being completed on a concrete hull. Most of the major ship classification societies, arising from their role of granting certificates of fitness to concrete gravity structures, now possess the necessary background of concrete technology to deal with prestressed concrete hulls.

For insurance purposes, it is likely that towing certificates will be needed for the tow over the high seas of any major concrete structure, whatever may be its intended mode of operation. For this purpose, requirements for damage stability will be specified, inner compartment walls being designed for full hydrostatic pressure and pipework valves and pumps being installed in the substructure.

The question of manning a structure during a tow raises further problems, for provision for a crew to monitor tow lines, lighting and stability can become expensive. The need for standby power, cooking and recreation facilities, fuel and provision storage, escape and emergency equipment, lifeboats, firefighting equipment, radio etc. for the two or three month tow serves no further purpose during the operational life of the structure. A ready solution lies in the installation of a complete crew accommodation package including all necessary equipment, which can be removed at the end of a tow and brought back for reuse. A typical prefabricated unit for a crew not exceeding four men can be hired for about £850/day (Fig. 6).

The tow from Western European construction yards to locations such as the Arabian Gulf is well within present-day experience. Typical structures for process plants are small compared with concrete oil production platforms with all-up weights in the range 15 000–30 000 t, and draught requirements seldom exceeding 5 m. Consequently, fleets of ocean-going tugs, as marshalled for North Sea tows, are not required and tugs with 11 000 ihp provide acceptable towing speeds. The planned tow to the Middle East from Northwest Europe is seldom longer than 8–10 weeks, and use can be made of the Suez Canal for structures up to 50 m wide without difficulty, although the passage through the canal may require supplementary tugs to maintain acceptable speeds. Alternative routes round the Cape of Good Hope may present the risk of more severe weather and affect the design, and also add some weeks to the voyage.

INSTALLATION

Concrete platforms for industrial plants can be designed for operational modes as permanently floating units, or alternatively as beached or grounded units at shore or riverside locations.

For permanently floating units, design studies need to cover the wave spectrum likely during the structure's life, and the response of the structure to the different sea states can be more critical depending upon the nature of the process carried out upon it. Even with strict requirements dictated by the plant, however, it is seldom that the behaviour as a whole needs special attention, and suitable solutions may be found by isolating those parts of the machinery that are critical, a normal procedure with any large undertaking. However, the mooring of the structure will require careful study and large-scale movement of anchors and mooring lines will generally be unacceptable. Constant survey of the moorings will then be necessary and periodic surveys of the concrete hull itself. In the design, careful thought should be given to enable these periodic surveys to be carried out without difficulty and provision made for replacement of the mooring lines if necessary. For the normal role, where a plant is beached upon soft estuarial soils or where a shore or riverside basin is excavated and the platform installed therein, the platform will remain unaffected by environmental forces and behave like any normal land-based structure. The cellular concrete raft has immense strength and stiffness so that weak alluvial soils can readily support the structure without the need for piling.

Installation in a land-locked basin will require careful preparation, and it is generally necessary to construct a number of pre-levelled concrete pads in the bottom of the excavation for the structure to settle upon. These can be constructed and accurately levelled at points where the structure is sufficiently stiff. The total load on these pads is only the change of buoyancy arising from the tidal variation, and is consequently much less than the total structural weight. Provision can be made for grouting with suitable materials any cavities beneath the base slab to ensure proper load distribution, before the basin is backfilled and the plant becomes surrounded by land. For economic reasons, a large plant can often be more economically built on a series of platforms of 4000–5000 m^2 area with hook-ups between the major process areas. When they are to be grounded, reasonable accuracy of location can be relied upon between individual units, and a tolerance of 10 cm in location can be maintained. Gaps between individual units of 2·5–3 m will provide space for the necessary pipe or channel links and the in situ work will be of little importance.

MAINTENANCE AND REPAIR

Prestressed concrete has proved to be the ideal material for structures exposed to a marine environment, having suitable durability and requiring no protective coatings which need periodic replacement, cathodic protection or other treatment. Records of thin-walled concrete ships (with 75 mm walls) or harbour structures such as the Mulberry 'beetle' units (32 mm walls) show that after 30 years or more exposure, no reduction in material strength can be expected. But marine structures are always exposed to risk through accidents or collision with other vessels, and it is here that concrete has advantages.

The repair of concrete structures can be carried out simply, even under water, by removal of all loose debris and making good with fresh concrete, using indigenous materials and no specially skilled labour force. The repair of damaged concrete platforms in the Gulf of Louisiana area has been reported to have been carried out without difficulty and repaired structures have carried on a useful life. One concrete ship which experienced a series of mishaps was the 300 t coaster *Violette*, constructed at Faversham in 1919 with a 7·5 cm thick flat bottom to its 38 m long hull. This was driven by a storm in January 1921 against Southend pier, demolishing part of it and finally grounding on one of the

Fig. 7. Violette discharging cargo at London Docks

piles which pierced the hull. It was written off as a total loss, but was then repaired and some 54 years later was reported to be still working[6] (Fig. 7).

The resistance of prestressed concrete to fire is even better documented.[11] Depending upon the cover given to the prestressing tendons, prestressed concrete is not only non-combustible but provides great resistance to fire, and various codes give well-established rules for assessing fire resistance. An effective demonstration of the durability of concrete barges was reported in US papers[12] in descriptions of the Bikini Atoll nuclear bomb tests:

> 'In the Bikini test two concrete fuel barges and a concrete drydock survived the blast, which sank five ships and damaged at least three score. One barge, No. 2160, was badly charred when the bomb ignited cargo fuel oil stored on her, but she was apparently in good shape otherwise with her decks well above the water line. The remaining yard oiler and the drydock, both farther removed from the blast center, suffered no apparent damage. Oiler 2160 was moored only a hundred yards from the *Nevada*, the bullseye for the test. Today the oilers were being towed in for a closeup inspection by naval construction men. In this same target area the carrier *Independence* was all but destroyed by the blast and subsequent fires, the cruisers *Pensacola* and *Salt Lake City* emerged with smashed stacks and superstructure, and the battleships *Pennsylvania* and *Arkansas* were so wrecked above the waterline they would have been useless in a naval engagement.'

MARSA EL BREGA REFINERY

Long before the recent North Sea developments had established the acceptability of long ocean tows for massive concrete structures, the buoyant platform technique had been pioneered with the installation at Marsa el Brega, Libya, of a refinery for Société Esso-Standard.[1] This structure, with a planned capacity of 1000 t/day of petroleum spirit, was constructed at Tamise, Belgium in 1962 and towed the 4500 km journey to its destination, where it is operating today.

The floating platform was 55 m by 24 m in plan and 5·40 m deep with a loaded weight of approximately 5000 t. The structure, of prestressed concrete, was divided by five longitudinal bulkheads and nine transverse bulkheads into 60 main compartments, the two foremost rows of bulkheads being divided again to stiffen the fore end of the structure to combat local high pressures due to wave slamming. The resultant 72 compartments were grouped into 13 zones for damage control and each zone was designed for full external hydrostatic pressure. Since the wave loads induce hogging and sagging movements of equal sizes, a uniform prestress was adopted.

The conventional calculation was made in which the hull was considered to be in equilibrium either on a crest or on a trough of a trochoidal wave of length equal to the hull length. This method, using an idealized wave and ignoring inertia, has proved satisfactory for a wide range of ships. The pontoon form is only distantly related to ship forms, so to reduce the effect of this unknown factor and to reduce towing resistance it was proposed to fit a steel ship bow, and the calculations were made for this design. When these were nearly complete, the steel bow was abandoned for reasons of cost, but the loading diagrams were retained; these were now conservative, giving an added factor of safety to compensate for the loss of the steel bow.

In choosing a design wave height the exceptional character of the journey was considered, but it was recognized that the pontoon could not put rapidly into shelter in bad weather. The design wave chosen had a height of 4·20 m and a length of 60 m. Calculated wave forces were increased by 50% to take account of the unknown still water loads.

The hydrostatic loads acting on the bottom and on the walls in the central region were easily evaluated. Above the water-line the loads, mainly dynamic from wave action, were more difficult to evaluate, and for the base slab, particularly at the fore end for about 20% of the length, the pressures due to slamming were assumed as 50 t/m^2.

The natural periods of roll and pitch were calculated as: pitch, 4 s; roll, 4·5–5·5 s. The maximum angle of roll was taken as the maximum slope of a wave of length equal to twice the beam of the pontoon and H/L ratio equal to 1/20; i.e., 9°. Tangential accelerations were proportional to amplitude and inversely proportional to the period of roll. From this, inertia forces were calculated for plant items.

The maximum working stresses in the concrete were taken to be 16 N/mm^2 (compression) and 1·6 N/mm^2 (tension). The pontoon was designed for no tension in the bottom slab or in external walls and prestress was provided in all three principal directions.

The 28 day compressive strength for 200 mm concrete cubes was 56 N/mm^2 (the maximum permitted by Standard NBM 15 for prestressed concrete). The cement content was 400 kg/m^3 and the water/cement ratio 0·50.

The base slab and external walls were concreted in one pour without construction joints, a delay of 8 h between joints being accepted. The prestressing tendons were 12/7 mm wires in 60 mm dia. ducts, and were anchored by bond in the lower part of the vertical tendons and similarly in the base slab.

Initially, half the transverse cables were tensioned. The longitudinal cables were then tensioned in four equal groups, the tensioning being equally divided between the two ends. Finally the remaining transverse cables were tensioned.

After grouting, tests for watertightness were made. Finally the surfaces in contact with water were painted with Fluoroc D which reacts with free lime to form a sealing film extending for a small depth into the concrete surface.

The total weight was greater than envisaged, giving a final draught of 3·75 m and a corresponding freeboard of 1·6 m. This proved inadequate. Waves often swept over the deck, frequently with violence, and it would appear necessary to specify a minimum freeboard of 2·5 m in future. During the voyage a storm was encountered in the Bay of Biscay during which the tow lines parted and the pontoon drifted in winds of force

Table 1. Approximate breakdown of overall construction costs (%) for a typical process plant

	Construction on site in Arabian Gulf	Construction on concrete platform
Civil construction costs	10	15
Towing costs, including insurance and certification	0	2
Platform installation	0	1
Process plant design and procurement	35	35
Process plant erection	40	20
Handling, shipping, insurance	10	0
Testing and commissioning	5	2
Final cost to client	100	75

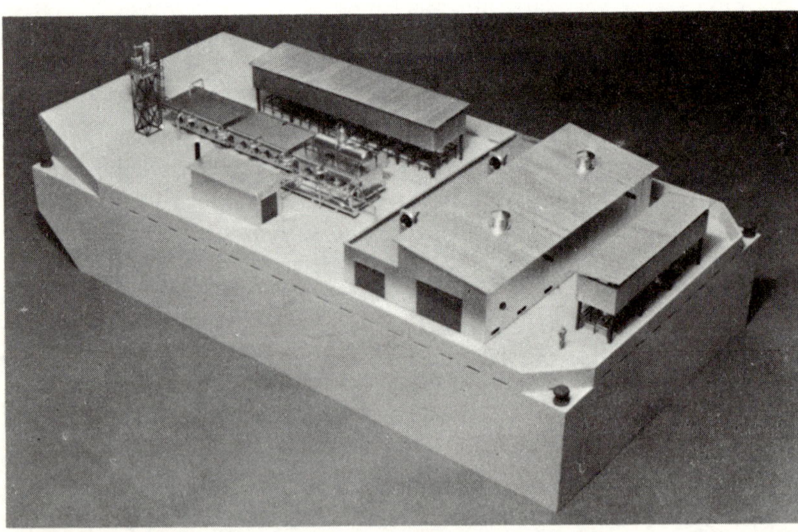

Fig. 8. Model of 600 t/day multiple-stage flash desalination plant

10–11. Maximum angles of roll and pitch of 14° and 10° respectively were observed (i.e., about 60% greater than anticipated) but no damage to plant was incurred.

ECONOMICS OF APPLICATION

For medium-sized platforms, prestressed concrete has shown considerable savings (20–30%) over similar structures of steel for first cost, disregarding the additional benefits of durability, fire resistance, lack of need for additional foundations, etc. To add to the prime cost figure, a capitalized figure for long term maintenance would indicate considerably greater savings. But the major economic advantage in using prestressed concrete platforms lies in the concept of the construction and commissioning of the complete production unit in an area where complete engineering facilities and skilled labour are readily available and labour costs are 30–35% of those of comparable expatriate labour. Proper coordination of civil works and engineering installations produces completed construction schedules considerably shorter than on-site construction and this programme saving does not need quantifying to produce capital savings of the order of 25%. The advantage of testing a complete unit and despatching it in complete running order has its obvious attractions when the location of operations is in one of the developing countries (Table 1).

The saving in time and overall capital cost can be seen to depend particularly on the degree to which existing engineering facilities and skilled labour can be employed in place of expatriate labour forces with the support facilities that they need. Consequently, simple civil engineering projects such as harbour caisson units can readily be constructed near the intended location and no economic advantage arises from construction overseas involving long ocean tows. However, as the plant engineering fraction increases, the viability of the concept readily becomes justified, and such facilities as desalination plants, water processing and injection plants etc., where the mechanical engineering may be 80% or more of the total figure, offer considerable economies.

Studies have been completed for such diverse projects as 600 t/day and 2000 t/day desalination plants (Fig. 8), 800 000 t/day water injection plants (Fig. 9), 200 t/day ammonia production plants, 40 MW electrical generation plants, a 20 000 t cold storage warehouse, a 260 room hotel (Fig. 10), a 270 000 barrel/day complete offshore refinery, and a 5000 person housing service unit (water, sewage and electricity).

In all, the concept for using prestressed concrete platforms for constructing a completed and tested facility for tow and installation for overseas has shown attractions of lower cost and shorter delivery, and prestressed concrete has been found to be the preferred construction medium.

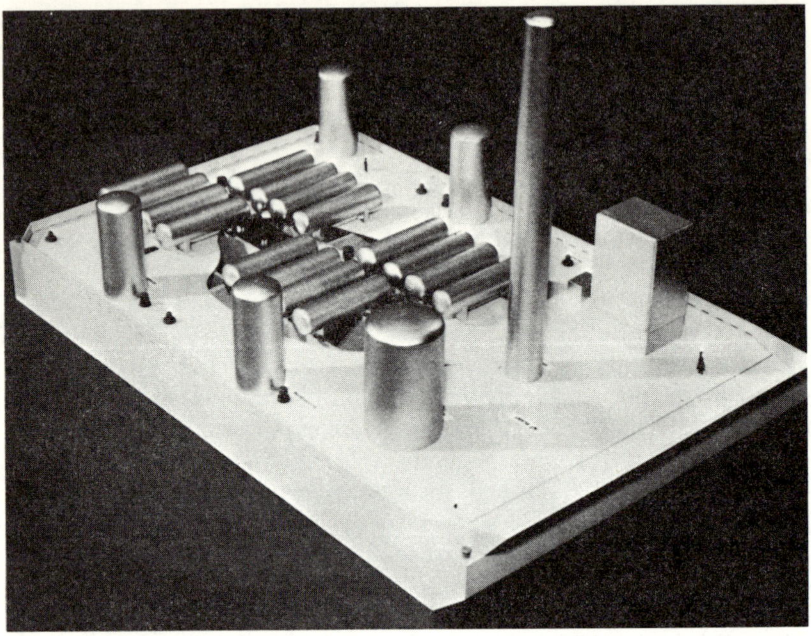

Fig. 9. Model of 800 000 t/day water treatment plant

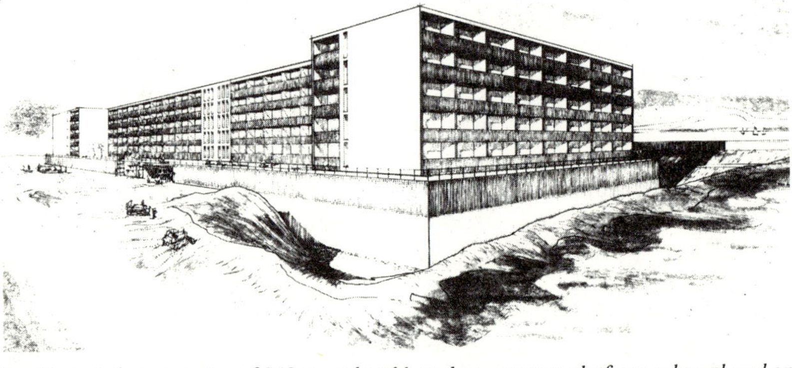

Fig. 10. Artist's impression of 260 room hotel based on concrete platform, when placed on location

ACKNOWLEDGEMENT

The author expresses his thanks to Sir Robert McAlpine and Sons Ltd for permission to publish this paper, and to his many colleagues for their assistance in producing it.

REFERENCES

1. Smet U. and Campus P. Une raffinerie flottante construite sur un ponton en béton précontrainte. *Bull. Tech. Bur. Veritas*, 1963, Sept., 158.
2. Mascaro F. and Jansky C. The world's first LPG offshore floating terminal. *Gastech 74: proceedings*. Gastech Exhibitions Ltd, Rickmansworth, 1975, 181.
3. Kosman W. Prospects are buoyant for floating plants. *Chem. Engng*, 1976, 26 Apr., 46.
4. Browne R. D. and Domone P. L. J. The long term performance of concrete in the marine environment. *Offshore structures*. Institution of Civil Engineers, London, 1975, 49–59.
5. Jellett J. H. The layout, assembly, and behaviour of the breakwaters at Arromanches harbour (Mulberry 'B'). *The civil engineer in war*. Institution of Civil Engineers, London, 1947, **2**, 291–312.
6. O'Driscoll P. The case of concrete coasters. *Concrete*, 1976, Aug., 14.
7. Fédération Internationale de la Précontrainte. *Recommendations for the design and construction of concrete sea structures*. FIP, Wexham Springs, Slough, 1974., 2nd edn.
8. Det Norske Veritas. *Rules for the design, construction and inspection of fixed offshore structures*. Det norske Veritas, Oslo, 1974.
9. British Standards Institution. *CP 110: the structural use of concrete*. BSI, London, 1972.
10. Bureau Veritas. *Note on the requirements for prestressed concrete ships*. Bureau Veritas, Paris, 1976.
11. Turner H. (Commission on Concrete Ships). Report on resistance of concrete hulls to fire and explosion. Fédération Internationale de la Précontrainte, Wexham Springs, Slough, in publication.
12. Concrete ships may be safe against atom bomb. *Scripps-Howard Press*, 1946, 3 July.
13. Paulling J. R. Ship structural loads for concrete ships and vessels. *Proceedings of the conference on concrete ships and floating structures*. University Extension, University of California, Berkeley, 1976.
14. Evans J. H. (ed.). Ship structural design concepts. Ship Structures Comm. project SR 200, 1974.

Concrete afloat. Thomas Telford Ltd, London, 1977, 43–50

The paper covers the historical evolution of ship classification to meet the needs of the industry, the development of rules and the creation of a field force. It also outlines the procedures involved in the assignment and maintenance of class.

The role of classification

C. F. ALGATE*

Historically, the older classification societies such as Lloyd's Register of Shipping came into existence to fulfil a need that was shared by all business interests connected with the operation of ships. Owners, underwriters, shippers and bankers accepted risks in varying degrees, and an independent technical opinion on the fitness of a ship to trade became a fundamental way of reducing some of these risks. Of all the parties concerned, the underwriters had the greatest total involvement in a venture and it is not surprising that it was they who took the first steps to 'classify' ships according to whether, from the construction, maintenance and manning points of view, they constituted good or bad risks.

In due course, the shipowners themselves recognized the advantages of complying with accepted standards, and took an active part with underwriters in the creation of a system of ship 'classification' based on defined technical requirements.

Despite the continuous evolution of ship technology since that time, the fundamental principles of classification remain: firstly, compliance with a standard of construction which ensures adequate strength, safety and reliability, and then adequate maintenance throughout the life of a ship to ensure a continuing standard of strength. Although manning was a factor at first considered by the underwriters, this is now a matter for owners and the government of the country of registration, rather than the classification society.

Classification as we know it today has grown step by step during the two centuries or more which have elapsed since the publication of the first known register of ships. Its creation was the direct result of a need for a basis upon which underwriters could assess a risk, whether in respect of ship or cargo. At the outset, ships were placed in categories dependent upon their general condition, the assessment of which did not always have a sound scientific basis. Nevertheless, it is true to say that ships were 'classified' in a manner which clearly indicated to subscribers whether they were considered to be 'good', 'middling' or 'bad' risks.

---

* Deputy Secretary and Controller of Classification, Lloyd's Register of Shipping, London.

DEVELOPMENT OF RULES

Largely because the basis for the assessment of class in the underwriters' register discriminated against ships built other than in the Thames, shipowners started their own register of ships at the end of the eighteenth century. Both registers suffered the consequences of what was virtually a 'cut-price' war, and this inevitably led to eventual amalgamation under the name Lloyd's Register of British and Foreign Shipping. In the protracted deliberations of joint committees prior to this we find the origin of today's fundamental principle that classification is assigned on the basis of clearly defined technical criteria. When the reconstituted society commenced operations in 1834, a uniform standard was adopted and appeared in the form of printed rules in the register itself. Since then, of course, the rules have been constantly changed to take account of developments resulting from service experience, new design techniques, materials etc. Thus we find the early 'Rules for the construction and classification of wood ships' being superseded first of all by rules for composite ships and then by the steel ship rules.

As far back as 1890, the General Committee of Lloyd's Register set up a technical committee for the expressed purpose of making recommendations for changes in the then existing rules or the adoption of new ones. Research was then in its very early stages, and the society collaborated closely with shipbuilders. The same collaboration exists today in that shipbuilders, as well as marine engineers, naval architects and other professional bodies, are represented on the technical committee.

Nowadays of course there is an increasing need for rules covering a variety of special types of ship. Lloyd's Register therefore publishes a whole range of separate rule books covering steel ships, liquefied gas carriers, floating docks, mobile offshore units, inland waterways vessels, yachts, refrigerating machinery etc., to which will shortly be added rules for concrete ships which are already in draft form. These will take account of the need for all concerned with concrete ships to consider how the continued structural strength of prestressed and reinforced concrete ships can be monitored economically during their service lives.

The rules for steel ships, which embrace detailed requirements for the design and survey during construction of essential machinery, electrical equipment, etc., as well as of the hull, contain specific provision for the acceptance of alternative arrangements provided these are considered by the committee to be equivalent to the society's requirements. This is of particular relevance following the introduction by the society of sophisticated computer techniques for the assessment of structural strength by direct calculation.

The basic symbol of classification is 100A1 with a Maltese cross (✠) prefix to indicate that construction has taken place under survey. The actual class assigned can, however, vary considerably, dependent upon type of ship and whether she is to be engaged upon unrestricted service or service within defined geographical limits. For example, we can have an ocean-going tanker which is classed 100A1 Tanker, or a tanker engaged on restricted service which is classed 100A1 Tanker for service between a and b. Within each category of restricted and unrestricted service there is no separate classification as such. The '100A1' symbol is appropriate in each case, not because the same standard of strength applies, but because each ship complies with the standard appropriate to its particular service.

For maintenance of the class to which a ship is entitled after compliance with the society's requirements for design and survey during construction, it is necessary for surveys to be held at periodic intervals throughout her life. The nature, extent and frequency of such surveys is described in the rules, which also require all repairs affecting the class of a ship to be carried out under survey. Where this is not possible because a surveyor is not available at the port concerned, the repairs must be examined by one of the society's surveyors at the first convenient port thereafter.

FIELD FORCE

To support the operation of the total classification system which is implicit in the application of the rules, and more recently, to provide a technical service in other important areas of engineering, Lloyd's Register has built up a world-wide organization. The present field force of surveyors exceeds 1700 and there is a large technical and administrative support staff in headquarters, where in former times nearly all plan approval work took place. As demand for the society's services spread, however, some of the conventional plan approval work was transferred to offices in major shipbuilding ports throughout the world, and this decentralization has developed to the point where, for ships of conventional type, comparable plan approval facilities using computer techniques are available in the major shipbuilding countries and in London headquarters.

The decentralization of plan approval in no way affects the application of unified standards which is an essential feature of a world-wide classification system. Just as identical standards and methods are applied in plan approval, so is uniformity achieved in surveys by the continuous updating of the surveyor's knowledge in terms of new technical developments and service experience with all types of ships and machinery. This is done not only by means of printed instructions and technical papers, but by specialized courses at the society's training establishments at Crawley and in Japan.

ASSIGNMENT OF CLASS

The shipowner's motives today in seeking to class his ship with a major classification society such as Lloyd's Register are probably little different from a century ago. If he is buying new, he naturally wants to ensure that he is getting value for money in terms of a sound ship which he can operate economically in his chosen trade. He knows that classification with a recognized society will assist him in arranging insurance and in fixing charters, and he is aware of the technical back-up which will be available to him during the life of the ship; or at least until he decides to sell her. Even then, it will not have escaped him that a ship with class maintained will be a more marketable commodity than one without class. This fact will be even more apparent to him if he is buying an existing ship for the first time.

Having decided that he wants a new ship built to Lloyd's Register class, the owner specifies this in his contract with the builder, and it is the latter who makes a formal request for the society's services and accepts responsibility for the payment of fees. It is also the builder who submits plans for approval as required by the rules, and who must comply with any amendments which the approval office considers to be necessary to conform with classification requirements. He must also ensure that all structural steelwork used in the construction of the ship has been approved by the society's surveyors during manufacture and that essential machinery has been built under the society's survey. Throughout the production cycle, from prefabrication to final trials, he will have to ensure that the local Lloyd's Register surveyors are afforded whatever facilities they require to verify by regular inspections that construction is proceeding in accordance with the approved plans, and that the standard of workmanship is satisfactory. In the case of large modern yards, builders normally have their own quality control procedures. Where these have been approved by the society, they may complement the surveyors' general supervision of construction.

On satisfactory completion of acceptance trials, the surveyor will issue a certificate to the builder, stating that the ship has been built under survey in accordance with the society's requirements and that he will be recommending to the committee that she be classed ✠ 100A1 etc. He may also hand over certificates covering compliance with international conventions where the appropriate inspection work has been undertaken on behalf of the government of the country in which the ship is to be registered. (A ship built to class will satisfy the structural requirements of the 1960 International Convention for the Safety of Life at Sea and the 1966 International Loadline Convention.)

A detailed report covering the classification survey is then sent to London accompanied by plans showing any deviations from the originally approved arrangements which the surveyor might have agreed with the builder. Provided the technical scrutiny which takes place when the report and plans reach headquarters shows that all classification requirements have been complied with, the surveyor's recommendation to the committee regarding assignment of class is confirmed, an official classification certificate is sent to the builder for handing to the owner in compliance with their contract, and details of the class are published in the society's register book.

PERIODICAL SURVEYS

As has been said earlier, the maintenance of class during a ship's service life is dependent upon periodic inspection as defined in the rules. The principle is that a major or 'special' survey of hull and machinery is held every 4–5 years. This involves complete opening up of all parts for inspection and the carrying out of recommended repairs. Because this can be very time-consuming, and therefore expensive, owners may elect to spread the survey over a period of five years. This is acceptable, provided all items are seen in rotation, and not more than five years elapses between inspections of individual items. The hull and/or machinery are then said to be on 'continuous survey'. Whether on a basis of special or continuous surveys, a ship also has to undergo minor or 'intermediate' surveys such as annual surveys of hull (which may be held afloat), drydocking surveys and surveys of tailshafts, boilers and other pressure vessels. The rules say that it is the owner's responsibility to see that all these surveys are held when due, regardless of whether notification has been received. Nevertheless, the society recognizes that even if an owner has his own planned maintenance scheme, regular advice regarding the current classification position of his ship or fleet can be of great value.

DATA PROCESSING

For many years, advice of surveys falling due had been sent on a limited basis, but it was not until detailed survey records were transferred to a computer that it became possible to extend this service to the form that exists today. In every ship, there are some 300 separate items of hull and machinery which are subject to special or continuous survey, quite apart from the boilers, steam generators, exhaust gas economizers, tailshafts etc., which may be due for survey at more frequent intervals. Each of these items or groups of items is allotted its own unique code number, which is recorded in a computer file along with the date of its last survey. The ground rules regarding the incidence of each type of survey for various ship and machinery characteristics form an integral part of programs which cause the survey records of about 11 000 classed ships to be scanned at prescribed intervals. The output from these program runs takes various forms, the most important from the owner's point of view being a listing at quarterly intervals showing all main survey dates,

Quarterly Listing — LLOYD'S REGISTER OF SHIPPING — PAGE 1

LR. NO.	SHIP'S NAME	GROSS TONS	DATE OF BUILD	FLAG
XXXXXXX	XXXXXXXXX	XXXXX	XXXXX	XXXXXXXXX

PLEASE TREAT THIS QUARTERLY LISTING AS A NOTICE OF SURVEY(S) BECOMING DUE AND/OR OVERDUE. SEE NOTE AT FOOT OF PAGE.

EXPIRY DATES OF STATUTORY CERTIFICATES

LOADLINE	30/06/74
SAFCON	30/06/74
SAFETY EQUIP.	01/01/76
SAFETY RADIO	03/02/75

A.N. OTHER SHIPPING COMPANY LTD.,
MARINE BUILDING,
MAIN STREET,
CITY,
COUNTRY

ML/NO	MAIN SURVEY DATES AND CONTROL	ASSIGNED DATE	DUE DATE	O/DUE	POSTPONEMENT NEW DATE	DUE NEXT 6 MONTHS
	CONTINUOUS SURVEY HULL	12-68	12-73	YES	06-74 APPROVED	
	CONTINUOUS SURVEY MACHINERY	12-68	12-73	YES	06-74 APPROVED	
	PERIODICAL LOADLINE INSP	25/02/73				
	ANNUAL CLASSIFICATION	02-73	02-74	YES	06-74 APPROVED	
	DOCKING	06-72	06-74			YES
1296	TAILSHAFT	06-72	06-76			
1346	NO.1 LOWER AUX. BOILER	02-74	02-76			
1410	NO.2 UPPER AUX. BOILER	09-73	09-75			
1489	NO.1 EXHAUST GAS ECONOMISER PARTLY HELD - ITEMS SEEN	07-72	07-74			YES
1515	INT. & EXT. EXAM. WITH MOUNTINGS & SAFETY VALVES					
1553	STEAMPIPES	12-68	12-73	YES	06-74 APPROVED	

ML/NO	CONTINUOUS MACHINERY SURVEY CONTROL	
***	ITEMS CREDITED SINCE LAST LIST ***	DATE
	INLINE TRUNK ENGINE	
0451	NO.6 CYL, COVER, PISTON, CONN ROD & BEARING, VALVES & GEARS	03-74
0933	STARBOARD INBOARD S.W. COOLING PUMP	02-74
1003	DIESEL OIL TRANSFER PUMP	03-74
0652	EMERGENCY AIR COMPRESSOR SAFETY DEVICES & ENGINE	03-74
0941	OUTBOARD MAIN ENGINE F W COOLER	02-74
0974	PORT INBOARD AUX ENGINE IN ENTIRETY	02-74
***	ITEMS OVERDUE ***	DUE DATE
	INLINE TRUNK ENGINE	
0201	NO.1 MAIN JOURNAL AND BEARING (P/P APPROVED 06-74)	12-73
0218	NO.2 MAIN JOURNAL AND BEARING (P/P APPROVED 06-74)	12-73
0226	NO.3 MAIN JOURNAL AND BEARING (P/P APPROVED 06-74)	12-73
0234	NO.4 MAIN JOURNAL AND BEARING (P/P APPROVED 06-74)	12-73

*ML/NO	CONTINUOUS HULL SURVEY CONTROL	
* ***	ITEMS CREDITED SINCE LAST LIST ***	DATE
*1841	NO.1 HOLD	02-74
*2006	ENGINE ROOM	03-74
* ***	ITEMS OVERDUE ***	DUE DATE
*1702	FORECASTLE DECK PLATING (P/P APPROVED 06-74)	12-73
*1865	NO.3 HOLD (P/P APPROVED 06-74)	12-73
*1897	NO.5 HOLD	02-74

0330	TV DAMPER OR DETUNER (P/P APPROVED 06-74)	01-74	*
0355	FORWARD SCAVENGE BLOWER (P/P APPROVED 06-74)	02-74	*
0564	INTERMEDIATE SHAFTING (P/P APPROVED 06-74)	03-74	*
			*

CONTINUED

NOTICE OF SURVEYS (See Chapter B of Society's Rules)
It is the responsibility of Owners to ensure that all surveys necessary for the maintenance of class are carried out at the proper time under the supervision of the Society's Surveyors.
The Society will give timely notice to Owners about forthcoming surveys by means of a letter or a Quarterly Computer print-out. Such notice does not absolve Owners from their responsibility to comply with the Society's survey requirements for the maintenance of class.

Quarterly Listing — LLOYD'S REGISTER OF SHIPPING — PAGE 2

LR. NO.	SHIP'S NAME	GROSS TONS	DATE OF BUILD	FLAG
XXXXXXX	XXXXXXXXX	XXXXX	XXXXX	XXXXXXXXX

PLEASE TREAT THIS QUARTERLY LISTING AS A NOTICE OF SURVEY(S) BECOMING DUE AND/OR OVERDUE. SEE NOTE AT FOOT OF PAGE.

ML/NO	CONTINUOUS MACHINERY SURVEY CONTROL		*ML/NO	CONTINUOUS HULL SURVEY CONTROL	
			*		
***	ITEMS DUE IN NEXT 6 MONTHS ***	TO BE EXAMINED BY	* ***	ITEMS DUE IN NEXT 6 MONTHS ***	TO BE EXAMINED BY
0572	PUMPING ARRANGEMENTS	06-74	*1915	NO.1 TWEENDECK SPACE	06-74
0398	PORT STEERING MACHINERY UNIT	06-74	*1922	UPPER TWEENDECK MID-SHIP SPACE	07-74
0371	FORWARD SCAVENGE AIR COOLER	06-74	*1807	NO.3 LOWER TWEENDECK SPACE	07-74
0800	FORWARD AUX FEED PUMP	06-74	*1815	NO.3 UPPER TWEENDECK SPACE	07-74
0817	AFTER AUX FEED PUMP	06-74	*2075	AFT PEAK TANK	08-74
1283	EXHAUST GAS ECONOMISER CIRCULATING PUMP	09-74	*	NO.3 PORT F W DOUBLE BOTTOM TANK	
	SEA CONNECTIONS		*2101	TEST	08-74
1146	PORT SIDE SUCTIONS	09-74	*	NO.4 CENTRE DOUBLE BOTTOM TANK	
1153	PORT SIDE DISCHARGES	09-74	*2132	TEST	09-74
			*3142	STEERING GEAR COMPARTMENT	09-74
			*3180	STEERING GEAR	09-74
			*2220	NO.1 PORT DEEP TANK	09-74
			*2251	NO.1 STARBOARD DEEP TANK	09-74

CONDITIONS OF CLASS (SEE KEY £ MEANS TO BE RENEWED)

01 ALL ME GUIDE BAR SUPPORT PTE LOWER ENDS EXCEPT NO.3 TO BE RADIUSED + WELDED REPS BY 05/74
02 NO.2 (S.F.) AND NO.5 (S.A.) GUIDE BAR SUPPORT PTES (CRACKED) * BY 05/74
03 UPPER HALF OF ME NO.4 CRANKPIN BEARG (BROKEN WHITE METAL) * BY 07-74

END OF LISTING

Fig 1. Example of a quarterly listing

and postponements where granted, together with continuous survey items either overdue or falling due within the ensuing six months or longer if required, also items which have been temporarily repaired but require to be re-examined, or items which need to be otherwise dealt with by a certain date as a condition of class maintenance. (An example of a quarterly listing is shown in Fig. 1.) Output for the field surveyor takes the form of a monthly microfilm containing the survey status of all classed ships. This makes him largely independent of head office records when consulted by local superintendents or masters of ships calling at his port.

Collectively, these two features represent a significant enhancement of the previous service and are covered by the normal survey fees, but it is almost certain that future developments in the area of direct computer communications will permit further improvements to the standard ship classification information system, provided realistic cost levels can be maintained.

Concrete afloat. Thomas Telford Ltd, London, 1977, 51–70

The American Bureau of Shipping has over 50 years' experience of classing concrete vessels. This has shown that concrete ships can be extremely durable, safe vessels, and economically a viable alternative to steel ships; it has also shown that concrete ships present design problems not found in comparative steel vessels, and innovative methods are required to overcome them. Cellular and modular construction, with the associated analytical methods of flat- and folded-plate theory, two-way slabs, and pre- and post-tensioning, are being utilized. Revolutionary hull shapes have been investigated and analysed for making better use of the compressive strength properties of concrete in resisting hydrostatic pressure. For the simpler structures of barges, design review for classification is possible using the traditional manual methods of naval architecture and structural engineering. For designs of increased complexity, ship motion analysis and the methods of finite element are employed for a more refined analysis. Finite element analyses were particularly useful in regions suspected of having higher stress concentrations, such as tank saddles. The ship motion analyses were used to determine what effect the response characteristics of concrete vessels has on the dynamic components of hull bending moment and shear force. The survey requirements for concrete vessels are no more severe than for steel vessels, with underwater surveys in lieu of dry docking an acceptable procedure for certain types of vessel.

Classification of concrete ships: historical background and current practice

R. A. CAMPBELL, BSc(Eng)*
K. T. CHANG, BSc(Eng), MS(Eng)†
S. G. STIANSEN, BSc(Eng), LLB‡

Classification societies are an integral part of the international marine industry. The majority of vessels engaged in commerce are built to classification societies' 'rules'; i.e., standards. In recent years the classification societies have also become involved with marine structures such as mobile drilling units, moored storage vessels, and fixed and gravity type platforms. The designers, owners and builders of a project all derive benefits from the classification process. When concrete is used as a construction material for these structures the concrete industry will come into contact with classification societies. The purpose of this paper is to discuss the American Bureau of Shipping's experience with concrete construction and generally outline the steps for classing a new structure. Included is a description of computer analysis of floating structures with specific reference to the unique characteristics of a concrete vessel. This paper also presents some of the ABS's specific recommendations for

* Surveyor, Hull Dept, American Bureau of Shipping, New York.
† Senior Research Engineer, Research and Development Division, American Bureau of Shipping, New York.
‡ Assistant Vice President, Research and Development Division, American Bureau of Shipping, New York.

major areas of design criteria, construction, and service review for concrete construction.

HISTORY OF ABS INVOLVEMENT WITH CONCRETE VESSELS

The earliest recorded use of concrete as a shipbuilding material was in 1848 when J. L. Lambot built a ferrocement rowing boat. Throughout the late nineteenth century and up to the present, small vessels have been built following Lambot's technique. The Norwegian ship *Namsenfjord*, launched in 1917, was the first self-propelled ocean-going vessel built from reinforced concrete. The ABS became involved in reinforced concrete construction in 1917 when this material was developed for shipbuilding as part of the US war effort. Concrete was a desirable alternative building material because of a shortage of steel plates and shapes. The ABS attended the construction of ten vessels (see Table 1) in this programme and granted a classification of ✠ A1 Annual Survey Experimental, Coastwise and West Indies Service. This class designation limited operation to relatively sheltered waters and required an annual survey on dry dock. Wig[1,2] describes the design and construction of these vessels.

Table 1. World War I concrete vessels classed by the ABS

Vessel name	Year launched
Atlantus	1919
Cape Fear	1919
Sapona	1920
Latham	1920
Cuyamaca	1920
San Pasqual	1920
Palo Alto	1920
Peralta	1921
Dinsmore	1921
Moffitt	1921

Fig. 1. Concrete No. 7, a 350 ft oil barge

The vessels had a total of approximately 15 ship-years service as cargo vessels, with the *Cuyamaca* and the *San Pasqual* the most successful with four years' service each. Upon termination of trading service, several of the vessels served satisfactorily for many years as storage barges. Reports indicate that oil was stored in the hulls with apparently no detrimental effects.

The major reason for taking these vessels out of service was their poor dead weight/displacement ratio which made them uneconomical to operate in the excess tonnage market which occurred following the war. Structurally the vessels were susceptible to impact damage. Also leakage problems recurred where tank bulkheads intersected the shell. Durability of the concrete was not a problem.

Concrete ship construction was basically dormant for 20 years until World War II when, for many of the same reasons as during World War I, ships were again built of concrete. The willingness to use this material reflected favourably on the earlier experience. The ABS was very active in developing design criteria, testing materials, and overseeing construction for this effort.

Five standard vessel designs were developed, from which a total of 104 vessels were built.[3] One standard design was for a self-propelled vessel; the other four designs were barges. Twenty-four self-propelled vessels were built and classed ✠ A1 (for full ocean service), 27 barges (design B5BJ) were classed ✠ A1 Barge, River and Harbor Service, and eleven barges (design B7A1) were classed ✠ A1 Fuel Oil Barge. The remaining barges were built under supervision of the Bureau but were not classed since they went directly into armed forces service. Fig. 1 shows a typical vessel of these World War II series; the *Concrete No. 7*, a 350 ft oil barge.

The structural designs were a variation of traditional shipbuilding using extensive internal stiffening of the shell, with the exception of design B5BJ, which was a slab type structure with minimum internal stiffening. Slab construction resulted in thicker shell elements than other designs but construction was easier and faster. One vessel of the slab design was completely built in six and a half days. The builders found that transverse internal stiffening allowed easier pouring and consolidation of the concrete within the complicated forms than did longitudinal stiffening. The requirements for cement cover thickness over the steel reinforcement were minima of $\frac{7}{8}$ in. on the outside surfaces of the shell and bottom, $\frac{3}{4}$ in. on outside surfaces of the deck and $\frac{1}{2}$ in. on all inside surfaces. It was later found that the $\frac{1}{2}$ in. minimum on inside surfaces was not adequate and had inhibited a good cement–steel bond. The construction programme for these vessels utilized a lightweight concrete with density of 120 lb/ft^3. This resulted in dead weight/displacement ratios of 0·50–0·53. Part (a) of Fig. 2 shows the shell details for design B5BJ, the slab type structure,

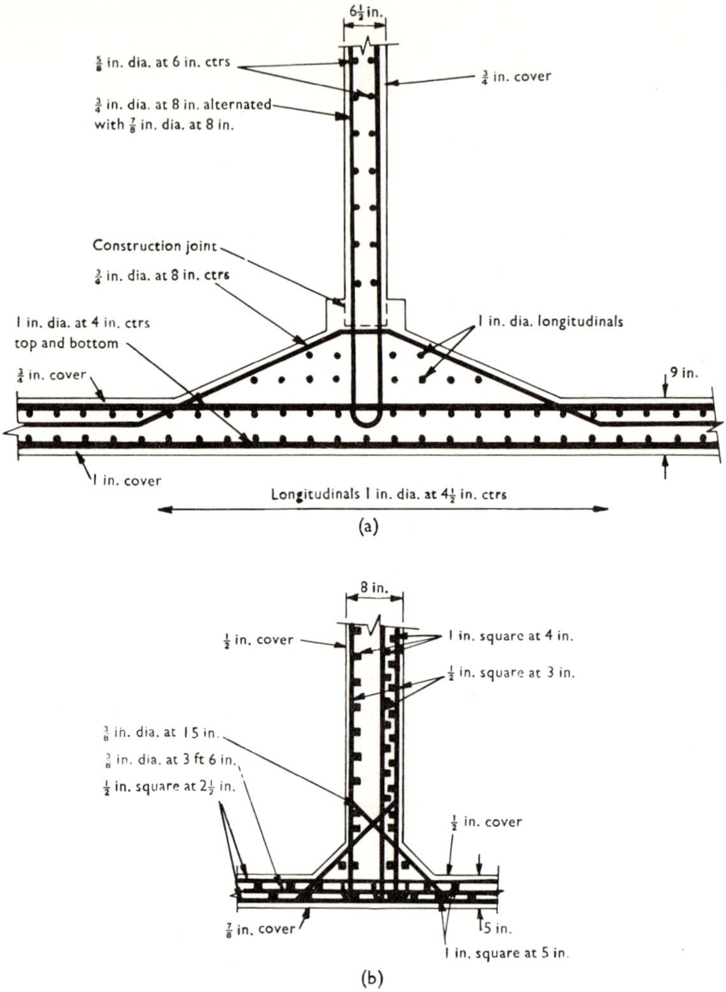

Fig. 2. Shell and bulkhead detail: (a) slab type structure; (b) stiffened plate type structure

and part (b) shows the same detail for the typical stiffened plate design.

Perhaps the most interesting structural detail of these designs was the treatment of the major construction joints. The ships were constructed in three pours: the bottom shell to the upper turn of the bilge; from the upper turn of the bilge to the deck joint; and finally the deck slab. This resulted in four major longitudinal construction joints at points where longitudinal stresses could be maximum. The construction procedure for these joints required that the dry cement be chipped and swept clean, moistened, and 'buttered' with a mortar mix of one part cement to one part fine sand. The new pour was made immediately on the prepared joint. This procedure was apparently adequate as no leakage or unusual problems were reported at the joint locations while the vessels were in service. Another design detail was to provide continuous longitudinal steel structure by butt welding the deck and bottom reinforcing rods rather than lapping them as was done elsewhere.

Most of the self-propelled vessels in this group were maintained in class for approximately four years. (Virtually all the barges constructed went directly into government service and thus were not kept in class.) The self-propelled vessels performed much better than their World War I counterparts. There are numerous reports of these vessels grounding with no damage other than a scraped bottom. The opinion of ABS field surveyors was that the ships were 'tender' to local impact but quite resistant to impact over large areas. The most common problem with the vessels was leakage through deep-tank bulkheads at the intersection with the shell. There were no significant problems of leakage through the shell, and surveyors often reported 'dusty' bottoms for the dry cargo vessels. After leaving trading service these vessels were used as storage barges, breakwaters etc. Like the World War I ships, the vessels served well in these functions, showing hardly any effects of corrosion on the reinforcing bars or any harmful deterioration of the concrete.

The absence of catastrophic, or ultimate, failure due to wave bending is the most striking positive result of concrete vessels' war service record. This is a significant accomplishment as the vessels were of considerable size, no prestressing techniques were used, the technology and construction materials were not as advanced as today's, and the vessels, particularly the World War II group, did experience rough ocean conditions. Concrete construction provides a very practical advantage over steel with regard to ultimate failure since the concrete will exhibit progressive deterioration in the form of considerable cracking before the failure occurs. The cracking will not propagate through the structure and, should the high loads be reduced, the cracks will close and can heal themselves. This cracking before failure allows the captain a period to take remedial action if significant structural deterioration begins during severe storm conditions.

Despite the history of satisfactory service, little activity in concrete ship construction followed the war, probably because of the relatively poor dead weight/displacement ratios of all operating concrete ships. In the late 1960s ferrocement construction for small vessels became more common and a series of 2000 ton prestressed concrete barges was built. These approaches recognized the weight penalty of reinforced construction and attempted to better utilize the included steel to gain a lighter structure. Ferrocement has been fairly successful in this regard and can result in a hull weight equal to that obtained with the more traditional building materials. The prestressed concrete barges resulted in dead weight/displacement ratios of 0·65, a considerable improvement over the earlier vessels. Two of the most recent concrete projects the ABS has classed, the *Ardjuna Sakti* and the *Cenec 20*, are floating moored structures and light ship weight was not a critical design parameter. The design factor which did suggest the use of concrete was the need for long term durability with limited maintenance.

The *Cenec 20* is a 118 ft reinforced concrete barge designed and built by Christiani–Nielsen Engenheiros e Construtores S/A, Brazil, for use as a floating concrete mixing facility. The intended operational area is in protected waters, thus the structure was designed for still water bending moments, local cargo loads, and local hydrostatic loads only. The standard Brazilian design code was applied directly, with no special safety factors included due to the water-borne aspect of the structure.

The *Ardjuna Sakti* is a floating storage facility for liquefied petroleum gas, built of prestressed concrete by Concrete Technology Corporation, Washington, for Atlantic Richfield Company. The designers, ABAM Engineers, utilized the high compressive strength of concrete and eliminated much of the secondary stress by using an unstiffened cylindrical shell type bottom spanning transversely between longitudinal supports. At the transverse bulkhead and tank saddles extra local reinforcing was included to resist the local combination of tertiary and primary stress. Fig. 3 shows the cross section of the vessel at a tank saddle. The vessel dimensions are: length, 460 ft; beam, 136 ft; depth, 56 ft; and displacement, 65 000 tons.

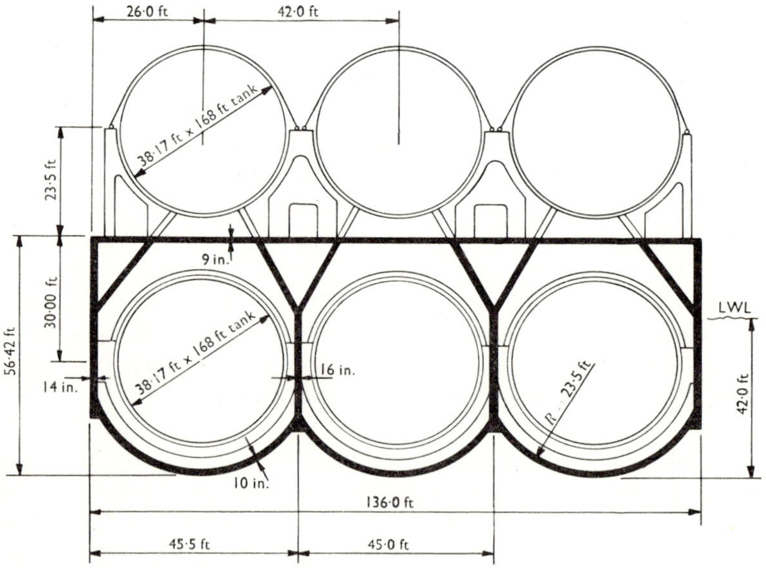

Fig. 3. Typical transverse section of Ardjuna Sakti

Although large concrete vessels were first classed over 50 years ago, no vessel has remained in class long enough to provide a normal lifetime record. The experience available indicates that a properly engineered and built concrete vessel will not experience catastrophic failure. Experience also indicates that concrete vessels are very durable in the sea water environment, but this conclusion should be tempered with an understanding of the structures built, which have been predominantly thick-shelled. As designers work toward higher dead weight/displacement ratios there will be proposals to use thinner shells. The experience with the durability of thick shells does not necessarily prove that thin shells will have as favourable characteristics. Accordingly, a conservative approach to serviceability criteria, which are predominantly crack control criteria, is justified until operational experience with thin-shelled vessels is available for evaluation.

CURRENT CLASSIFICATION PROCEDURES

The ABS is prepared to accept concrete vessels for unlimited classification. These vessels are to be built to the specifications of the concrete industry and the ABS. The Bureau anticipates about a 40 year life for concrete vessels and expects designers to address the problems of endurance as well as of ultimate strength. Following is an outline of the classification procedure for a new design with specific recommendations for detail design. This outline is typical for class societies but the specific recommendations are only those of the ABS.

First phase of classification

The first step is the submission of design data to the technical staff. The classification society and designer should agree on design methods, loads, and factors of safety before any engineering work is done. A society's position on these matters is generally available in a published document, but when dealing with new materials it is advisable to request specific confirmation on each proposed project. Having established the design basis, the designer can proceed with confidence that his methods will be acceptable to the society. When submitting the design for approval, adequate lead time is necessary so that the society can respond before any unchangeable decisions are made. The information submitted should include the building specification, construction plans, engineering analysis and quality control specification.

Building specification. This document should include the materials specifications, mix specifications, steel and cement placement procedures, stressing procedures, formwork and curing procedures, and alignment tolerances. Classification societies will generally accept the established guides of the concrete industries on these items, with specific emphasis given to producing cement which is watertight and resistant to a salt water environment.

Construction plans. A complete set of construction plans should be submitted to the classification society for review. These plans need not be as detailed as field plans but should show all steel placement.

The ABS requires that the midship section scantlings affecting longitudinal strength (wall thickness, steel placement and content) extend for 40% of the midship length. Outside this area these scantlings may be reduced.

The ABS also requires that for post-tension construction the design incorporate 5% excess post-tensioning capability. This may be provided by extra ducts which can be filled with grout if unused or by using a 95% factor for calculating effective prestress. This factor is required to account for post-tensioning not obtained due to blocked ducts or slipped anchors.

Engineering analysis. The designer should submit his complete calculations detailing the codes he followed, the design loads and the results. If standard computer programs are used, the programs should be named and a listing of the inputs and outputs submitted. If non-standard computer programs are used, a complete description of the program should be provided, together with the above information. The classification society will review the designer's submitted analysis as well as make independent studies as it considers necessary. The 'rule' requirements as published by the societies are not always adequate to determine the scantlings for a specific design, especially for vessels built of unusual materials. Under such circumstances a more rigorous review using computer analysis is necessary. For current practice, the two-dimensional or three-dimensional finite element computer analysis approach is generally adopted to examine the stresses in longitudinal and transverse structures. In the two-dimensional analysis, portions of the ship structure that are of particular interest are isolated from the main structure and are represented by mathematical models. The proper choice of elements, boundary conditions and loading is essential if the model is to accurately resemble the actual structure. If the structure cannot be adequately modelled by two-dimensional methods, a more sophisticated three-dimensional analysis should be used.

For concrete structures the available computer programs generally use a linear elastic analysis to calculate forces and moments of the members in the overall structure. These results then can be used to design or check the steel reinforcements in structural members. The general purpose finite element programs NASTRAN, ICES STRUDL-II, MARC etc. are often used to analyse concrete structures. In addition a separate reinforced concrete version of ICES STRUDL-II and a series of programs maintained by the Portland Cement Association provide features which aid in the design or checking of reinforced concrete structures. The basis of these programs is the working or ultimate strength requirement of the American Concrete Institute.

The ABS has developed the ABS/DAISY computer system for two and three-dimensional analysis by the finite element displacement method. This computer program system has been extensively employed in the design review and analysis of large tankers, high-speed container ships, liquefied natural gas vessels, general cargo ships, barges and bulk carriers, with structures ranging in size from local detailing to an entire ship hull.

The generation of finite element models such as complex ship structures generally requires a large expenditure of manpower and effort to prepare and verify input data. The ABS has developed a series of programs to effectively reduce the effort and time needed to generate structural models. A system of preprocessor programs essentially designed for ship structure has been developed as part of the DAISY system to analyse the complete three-dimensional hull structure or parts thereof. The two-dimensional preprocessor system is used to automatically generate planar structures such as web frames, girders and brackets. The postprocessor system consists of programs to display the DAISY calculated results, including printout of the entire results and/or selective output and plotting of model displacements and stress contours on an incremental plotter. To handle the analysis of reinforced concrete ships, the preprocessor and postprocessor in the DAISY system can be readily used to generate ship structural input data for above-mentioned commercially available computer programs capable of solving concrete structural problems.

As far as the determination of sea loads is concerned, the simple quasi-static approach is employed to calculate hull girder bending moments in the 'rules'. This approach involves the poising of a ship in equilibrium on a hogging or sagging standard wave profile. The wave is assumed to have the same length as the ship and a height which is expected to cause the maximum stress in the ship in actual service over a period of time. The effective wave height used in the 'rules' is based on operation experiences and the wave-induced loads which are determined by the ship motion long term prediction methods.

The basic information necessary for long term load prediction is the wave-induced responses in regular waves which are generally based on the strip method. The computer programs SCORES, developed at Oceanics, Inc. and sponsored by the Ship Structure Committee (report SSC-230), and SEAKEEPING, developed at the Massachusetts Institute of Technology, are examples of applications of this method. The two programs are similar in predicting ship responses but are somewhat different in other respects. SCORES can include surge motion approximated by the corresponding result of an ellipsoid. SEAKEEPING includes a special routine to handle bulb sections which cannot be properly treated by the commonly used 'Lewis' form. It also considers the non-linear rolling of the ship as well as the occurrences of slamming, deck wetness and propeller racing.

The prediction of ship responses in irregular waves (short term response) can be made on the basis of the results in regular waves and wave spectra of a particular ocean area. This is done under the assumption suggested by St Denis and Pierson[4] in 1953, that a vessel in irregular waves can be treated by the principle of superposition of responses to regular waves corresponding to a full range of wave spectrum. By then assuming a certain distribution of the short term response properties, the long term response can be predicted using probability methods. In general, the prediction procedures used by different classification societies are similar; that is, the probability is separated into two parts, one being the probability of the ship response exceeding a specified level, the other being related to the probability of the vessel encountering various sea conditions. However, differences exist in handling these two probability functions. As the basis for its program SHIPMOTION, the ABS expresses the response probability of exceedance as a normal distribution of narrow banded (Rayleigh) short term parameter within each wave height group (weather group) and the probability of sea condition on the basis of weather groups. This is in contrast to another approach wherein both probability functions are related to wave period groups. The ABS has made efforts to collect full-scale measured results to compare with

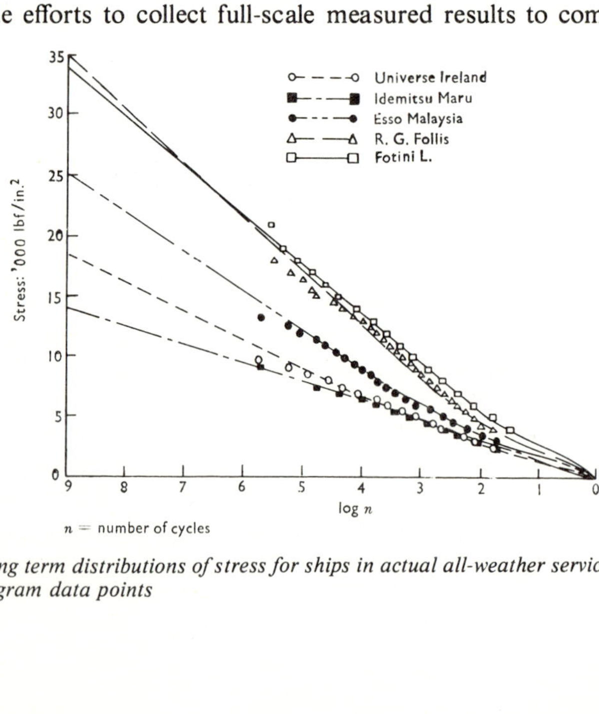

Fig. 4. Long term distributions of stress for ships in actual all-weather service, compared with histogram data points

the theoretical predictions of the SHIPMOTION program. Fig. 4, adapted from Little and Lewis,[5] illustrates an example of such a comparison. In this figure measurements from four tankers and one bulk carrier, represented by the data points, show very good agreement with the computed results, represented by the faired lines.

The availability of the analytic long term prediction method enables classification societies to develop an effective wave height which, when used as the design wave in a simple equilibrium wave bending calculation, will produce the same response as the long term result. The design waves published in societies' 'rules' represent a wave envelope based on the result of long term predictions for a series of typical vessels. Vessels designed with these values would not be limited in service since the North Atlantic wave spectra is used as the calculation basis. For vessels designed for limited service, using the North Atlantic wave heights would be conservative, since the vessel would be operating in less severe sea conditions.

For concrete ships the long term prediction technique mentioned above can still be applied. Concrete construction will result in geometric particulars and weight distributions which vary significantly from those of a conventional steel vessel, and therefore the effective wave heights published by classification societies for steel vessels would not necessarily be accurate for a concrete vessel. In addition, the service life of concrete ships, which may be much longer than that of steel ships, can be taken into account when the long term prediction method is applied. A ship motion analysis to determine design values should be carried out for concrete vessels greater than 200 ft in length.

The design review carried out for classing the vessel *Ardjuna Sakti* provides an example of the application of structural and ship motion computer programs. The design criteria, including loads, assumed material strengths, and factors of safety were agreed to by ABS technical staff and the designers before the actual design work was begun. The designers agreed to submit for review their engineering calculations as well as the usual construction plans. The Bureau's personnel made an independent structural analysis utilizing the ICES STRUDL-II program, and also reviewed the designers' calculations. Fig. 5(a) shows the computer plot of the finite element model representing a typical tank saddle as shown in Fig. 3. The model consists of beam and membrane plate elements. Since the transverse structure represents only a part of the entire three-dimensional vessel, elastic supports are located at the bottom to represent the stiffness of the surrounding longitudinal structure. Various loading conditions were analysed with respect to the structural model. Fig. 5(b) shows the plot of deflections due to fully loaded tanks, hydrostatic sea loads to the deck at side, and a hydrostatic deck load as required by the *ABS rules for building and classing steel vessels*. The

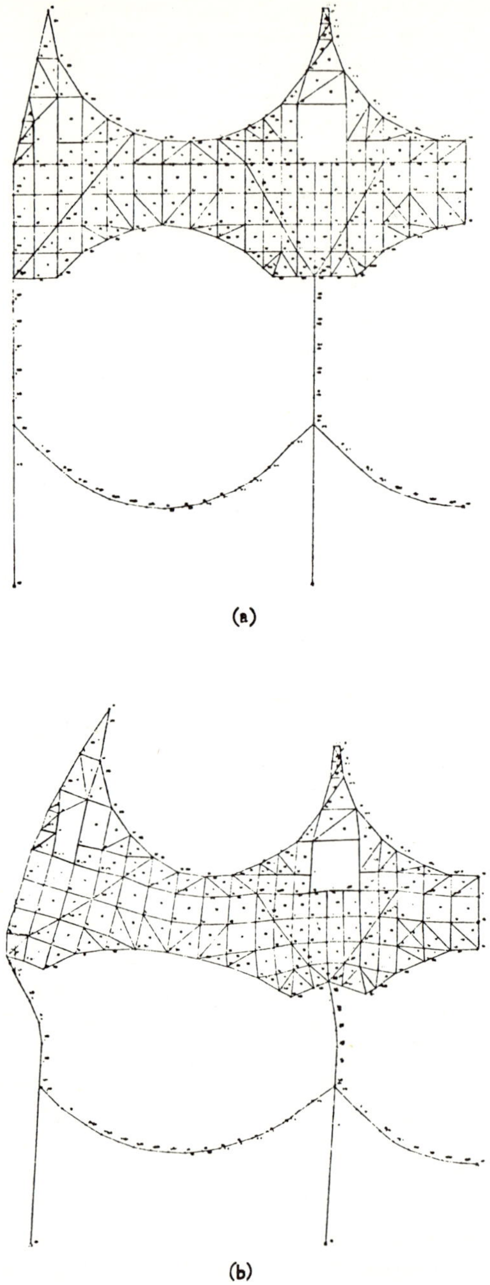

Fig. 5. *Computer plots: (a) two-dimensional finite element model of transverse tank saddle; (b) deformed shape of two-dimensional finite element model of transverse tank saddle*

dynamic effects are treated by a magnifying factor obtained from the ship motion analysis. The model is for one half the structure because of the symmetrical nature of the vessel.

A ship motion analysis was used to find acceleration factors for the cargo tanks and cargo handling equipment. The analysis was done with the ABS SHIPMOTION computer program which is regularly used for predicting dynamic behaviour of conventional ships. Two modes of operation were analysed, the ballast condition on the delivery voyage and the light ship condition at the vessel's moored location in the Java Sea. The wave spectra for the delivery voyage from the northern coast of Washington, USA to the Java Sea was developed by comparing samples of winter wave measurement at Station Papa in the Pacific to those at Station India in the North Atlantic. The comparison revealed that wave spectra shapes at these two stations are almost identical; hence the 80 wave spectra for ten significant wave heights for the North Atlantic, which is perhaps the most accurate wave data currently available, was used for this analysis. During the delivery period the dominant wind direction is from the south-east, thus the probability of wave direction with respect to the ship's heading is not evenly distributed. The wave spectra for the Java Sea was formulated from data in IMCOS Marine,[6] Hogben and Lumb,[7] Scott,[8] and Hoffman and Lewis.[9] The moored condition was modelled by assuming 75% of the waves were encountered head-on and the remaining 25% encountered on either side forward. The vertical and lateral accelerations at 14 points on the vessel were calculated. These accelerations were applied to the design of machinery, foundations and the tank saddles.

Quality control specification. A quality control procedure covering all facets of the construction should be submitted for the classification society's information and for the guidance of the society's field surveyors. The society will comment on these procedures where they are believed inadequate. The importance of quality control cannot be underestimated and the extent of this specification should reflect this importance.

The ABS will accept certificates of suppliers as proof of quality for the cement, aggregate and secondary steel reinforcement. Tendons for prestressing are to be tested and certified under the supervision of a surveyor. The builder should provide a table of building tolerances and a method of insuring that the tolerances are met.

The classification society will make an independent review of the above material using the codes of the concrete industry in conjunction with the society's own guide-lines. Of the submitted material the construction plans will be approved; the other material will generally be reviewed but only commented on. The society will amend the construction plans when the review reveals deficiencies.

Second phase of classification

The second phase of classification is field surveying during construction. A surveyor is assigned to a project for this purpose and will remain on the project from beginning to end. The surveyor's responsibility is to ensure that the vessel is built according to the specifications, the approved construction plans, and the quality control procedure. The technical staff is always available to advise the field surveyor when construction circumstances require deviations from the agreed plans. Builders and owners recognize that the field surveyor is working for their best interest and provide complete co-operation. For concrete construction, field surveying can be broken down into two broad areas: concrete quality control and overseeing construction.

Concrete quality control. Construction of a concrete vessel differs most fundamentally from that of a steel vessel in that the major building material, concrete, is produced at the construction site rather than supplied to the site already manufactured and, more importantly, tested for quality. A good quality construction job is highly dependent on the concrete and therefore the surveyor will give particular attention to the builder's procedures in this area. The surveyor will monitor the quality of all materials, require proper material storage, and monitor concrete mixing, testing and placing procedures.

The ABS requires that the concrete industry's standard slump and compression test procedures and schedules be followed and that the results be made available to the field surveyor. In addition, the Bureau urges that the quality control procedures specify an early strength test which can be correlated to the 28 day strength values.

Overseeing construction. Concrete construction will require more participation by the field surveyor in the actual construction process than does steel construction. The surveyor will inspect the formwork, steel placement before pouring, concrete pouring, joint preparation, prestressing operations, and generally all aspects of construction which will affect the strength of the structure. After the forms are removed from a completed section the surveyor will inspect for satisfactory tolerance control and concrete distribution. The surveyor will determine the necessity for rebuilding based on the seriousness of the building deficiencies.

The ABS suggests the following procedure for various construction problems.

Construction details should be carefully considered with regard to protecting the primary reinforcing from corrosion. Since adequate concrete cover affects corrosion, this will be closely monitored by the surveyors. For post-tensioned construction all tendons should be ducted, all ducts grouted, and the ducts continuous across construction joints. Where anchors for prestressing tendons are placed on exterior surfaces, special consideration for water protection is necessary.

Construction joints are particularly important in corrosion control and cargo containment since they are a discontinuity in the cement structure and have the greatest possibility for leakage. The following building and testing procedures are suggested for construction joints.

(*a*) For all dry-to-dry joints, both surfaces of the joint should be sandblasted, cleaned, and coated with an appropriate epoxy just before joining. Prestress should be applied across the joint while the epoxy cures.
(*b*) For all exterior dry-to-wet joints and interior dry-to-wet joints where the space is used for tanks, the dry surface of the joint should be sandblasted, cleaned, and coated with an appropriate epoxy just before concreting.
(*c*) For all other joints, the dry surface of the joint should be heavily sandblasted, cleaned, and in a 'damp' condition before concreting.
(*d*) All joints subjected to a hydrostatic head and all dry-to-dry joints exposed to the weather under normal operating conditions should be hydrostatic or vacuum tested depending on joint location. A hydrostatic test requires holding a water head equal to the design head on the joint. A vacuum test requires drawing a vacuum on one side of the joint and introducing an indicator solution on the opposite side which will be drawn through any leaks. At sites where these tests present unusual difficulties, equivalent tests may be considered.
(*e*) All dry-to-wet joints exposed to the weather under normal operating conditions and all joints not previously mentioned which are in watertight subdivision bulkheads should be air or water stream tested. This requires holding pressure on one side of the joint and providing means for detecting leakage through the joint.
(*f*) If leaks are found, the joint must be repaired and retested.

For prestressed construction a record of the stress actually obtained in each tendon must be kept. The designer should continually evaluate this information to determine if design modifications will be necessary to ensure the required final prestress. When post-tensioning is used, the prestressing force should be determined for each tendon by both measured elongation and jack gauge pressure. This information should be supplied to the society's field surveyor and technical staff.

All reinforcing steel which protrudes from an already cast element should be protected against corrosion and damage during construction. All post-tensioning ducts should be capped before use to prevent the entry of foreign water.

Since low slump concrete is used for construction, vibration will be required to insure good flow around the steel. The normal curing procedures may be followed, with the added provision that salt water spraying is not acceptable.

It is anticipated that concrete vessels will generally have greater weight than equivalent-size steel vessels. A concrete vessel is especially susceptible to damage from deflections during construction since it does not attain full strength immediately and design prestressing may not be applied until some time after the concrete is placed. Building ways must be adequately strong to support the vessel with no deflections during all phases of construction.

In the event that cracks occur in the structure during construction, the builder should consult the designer and the society's field surveyor and technical staff to determine what repair procedures will be necessary. The ABS will require repair of any cracks of width greater than 0·005 in. Certain cracks 0·005 in. and less may go unrepaired on the assumption that self-healing will occur.

Third phase of classification

The third phase of classification is in-service surveying. Through this in-service function the classification society remains involved with the vessel throughout its operating life. This involvement provides the society with a large amount of practical experience which forms part of the basis for rule development. The operating life for concrete vessels could easily approach 40 years.

The immediate purpose of in-service surveying is to ensure that a vessel remains in seaworthy condition. There are two aspects of this function: establishing a survey schedule which limits the time between inspections and a survey specification which details the items to be inspected.

Survey schedule. The final step in the classification process is the sea trials, which surveyors witness to ensure that all systems function under operational conditions. After a vessel is formally accepted for class it is placed on a regular schedule of classification surveys. The basic requirements of this schedule are for an annual survey and a biennial dry-docking survey for damage and obvious failures; and a special periodic survey every four years (or five years including a year of grace), which entails a more detailed inspection of the total vessel, including gaugings as the effects of corrosion become apparent. These basic requirements can be varied when special procedures are followed, and with the classification society's specific approval. In addition to the scheduled surveys, damage to the hull, machinery or equipment which affects or may affect seaworthiness or classification should be submitted by the owners or their representatives for examination by a surveyor.

Standard exceptions to the survey schedule for steel vessels are for drilling units and permanently moored storage facilities. In recognition of the problems of dry docking these vessels, in-place underwater surveys are accepted in lieu of all dry-docking requirements. The underwater surveys are done by qualified divers who clean a representative number of spots and examine them using television or video tape, with a surveyor, who is in direct communication with the divers, monitoring the closed circuit screen.

The possibility of altering the survey schedule for concrete vessels is complicated by the machinery survey requirements and the International Convention on Load Lines* requirements. The Convention on Load Lines requires an annual survey to ensure continued watertight integrity of the vessel, and a survey every five years to ensure continued structural adequacy of the vessel. The five-year survey requires dry docking. The classification society's machinery surveys are independent of those for the hull structure. The annual machinery survey is limited to a general inspection of visible machinery and does not require dry docking. Special periodical machinery surveys are required every four or five years during which, among other things, all openings to the sea are inspected. The propeller shaft of self-propelled vessels must be completely inspected every three or four years, depending on the shaft seal arrangement. Inspecting the openings to the sea and the shaft is most conveniently done on dry dock, but satisfactory procedures for fulfilling these requirements while the vessel is afloat can be used. All other machinery survey requirements can be fulfilled while afloat.

Although past experience with concrete vessels would indicate that dry-docking requirements could be relaxed since concrete is not corrosive, the classification societies are limited in flexibility by the Convention on Load Lines requirements as well as those of various nations' regulatory bodies and the practical problems associated with the machinery surveys.† The ABS proposes that concrete vessels follow the same basic survey requirements as steel vessels with the exception that the requirement for dry docking every two years be relaxed. Rather than a dry docking, an internal examination of the hull will be acceptable. Tank vessels carrying dangerous cargoes in integral tanks must be gas-freed before internal inspection. If an owner elects not to gas-free at this two-year interval, an underwater 'spot' survey will be required.

* A vessel's load line is the maximum draught at which it is allowed to operate. The load line is determined from the vessel's geometric particulars, stability criteria and the vessel's structural strength. The International Convention on Load Lines is an international agreement to ensure uniformity of load line assignments among maritime nations. Virtually all vessels trading in international waters require a load line. Each country may individually determine the loading limitations for vessels operating only in its coastal and inland waters.

† Concrete vessels will experience approximately the same amount of marine growth as steel vessels. Dry docking for cleaning the bottom is not a classification item.

The equivalent of steel gauging will be required for concrete vessels. This process should be a non-destructive inspection which will signify any internal deterioration of the structures. The need for gauging steel is normally determined by a visual inspection of the extent of corrosion. Concrete vessels will be 'gauged' when the shell shows excessive concrete deterioration, a visual inspection indicates reinforcing corrosion, or after the vessel has been in service for 16 years. Currently, ultrasonic techniques appear to be most advanced for this inspection.

Survey specification. The survey specification for concrete vessels is similar to that for steel vessels in terms of what should be inspected, but differs in terms of the type of deterioration expected. The surveyor will inspect the structure for cracks, superficial spalling of concrete, major concrete erosion, corroded reinforcement and structural damage. Generally, all defects will require repair, but some minor repairs may be postponed until the vessel is due for its next dry docking.

The necessity for immediate repairs is to the surveyor's discretion and is determined in part by the location of the damage. During the annual surveys any reported deterioration will be inspected and reported, thus providing a history of the problem.

Vessels which will be dry docked present no unusual repair problems. The concrete industry's patching and crack injection procedures are acceptable methods for repair. Vessels which are not dry docked will require special repair procedures. Repairs above the waterline will require especially careful surface preparation to ensure good bonding. Minor repairs under water can be accomplished using materials designed for this application. If the deterioration is of a more extensive nature, the repair area should be free of water during the repair. A very serious structural failure will probably cause the vessel to be taken out of service for repairs at a shipyard.

All repairs should be carried out by qualified personnel. The ABS will allow injection of cracks below the waterline from the vessel's interior where possible. Corroded reinforcement must be exposed and cleaned or effective redundant reinforcement provided.

CONCLUSION

The classification society's function is to serve the marine industry by providing guidance for building safe, durable structures. When a new material is proposed for construction purposes, the society must determine its policy by studying the material, consulting with experts who have used the material, and studying past experiences with the material. Based on this information, the society will establish its requirements to reflect its degree of confidence in the material.

The history of concrete for marine construction is very favourable.

There is little doubt that a well designed, well built concrete structure will have a longer service life than a comparable steel structure. In consideration of this, the ABS will accept concrete vessels for full, unrestricted classification. The procedures for classing a concrete vessel are slightly different from those for classing a steel vessel. Due to the novelty of the material, the designers and classification society should jointly determine the design criteria for a project based on the type of construction, properties of material, and intended service. Also, due to the novelty of the material and the structural arrangements devised to fully utilize its properties, computer techniques will probably be required for proper structural analysis. The dynamic responses of concrete ships will vary from those of a similar steel vessel and thus ship motion computer programs should be used to determine the motions, dynamic loads and effective design wave heights. Concrete construction will require greater on-site building inspection and monitoring of quality control procedures than steel construction but a slightly relaxed in-service survey schedule will be accepted.

Future design work will probably be aimed at improving the dead weight/displacement ratio of concrete vessels. The two approaches to this problem are the use of lightweight concrete or reduced scantlings due to higher strength materials or more critical design methods. Research will be necessary to determine the effect of these approaches on the cracking behaviour of concrete, which in turn will affect the durability of the structure.

Concrete is currently a practical material for gravity platforms and floating storage facilities. The ABS believes that with the confidence gained from increased service experience and the expected development of the technology of marine concrete construction, concrete will eventually become an attractive alternative to steel for transport vessels as well as fixed and moored facilities.

REFERENCES

1. Wig R. J. Present status of the concrete ship. *Trans. Soc. Nav. Archit. Mar. Engrs*, 1918, **26,** 185–198.
2. Wig R. J. Method of construction of concrete ships. *Trans. Soc. Nav. Archit. Mar. Engrs*, 1919, **27,** 1–28.
3. Anderson A. R. Prestressed concrete floating structures. Society of Naval Architects and Marine Engineers spring meeting, Vancouver, 1975.
4. St Denis M. and Pierson W. J. On the motions of ships in confused seas. *Trans. Soc. Nav. Archit. Mar. Engrs*, 1953, **61,** 280–357.
5. Little R. S. and Lewis E. V. A statistical study of wave-induced bending moments on large oceangoing tankers and bulk carriers. *Trans. Soc. Nav. Archit. Mar. Engrs*, 1971, **79.**
6. Wave data of Java Sea, Marsden square 325, 1956–1966, supplied by IMCOS Marine.
7. Hogben N. and Lumb F. E. *Ocean wave statistics*. National Physical Laboratory, Teddington, 1967.

8. SCOTT J. R. *A coastal spectrum for model tests*. Vickers Ltd, St Albans, 1966.
9. HOFFMAN D. and LEWIS E. V. *Wave spectra for design calculations, southwest corner of Java Sea*. Webb Institute, Glen Cove, NY, 1974.

BIBLIOGRAPHY

ALLEN R. T. L. and GREGORY-CULLEN J. *The inspection, maintenance, and repair of concrete offshore structures*. Ship and Marine Technology Requirements Board, Department of Industry, London, 1974.

CURRY R. *Prestressed concrete barges*. American Bureau of Shipping, New York, 1966.

Design and control of concrete mixtures. Portland Cement Association, Skokie, Ill, 1968.

Guidelines for the construction of ferro-cement vessels. American Bureau of Shipping, New York, 1969.

HAVILAND J. American concrete steamers of the First and Second World Wars. *Am. Neptune*, 1962, July.

International Conference on Load Lines, 1966, Final Act. Inter-Governmental Maritime Consultative Organization, London, 1966.

Inspection of concrete. *J. Am. Concr. Inst.*, 1975, June.

MOE J. Feasibility study of prestressed concrete tanker ships. *J. Am. Concr. Inst.*, 1974, Dec.

Recommendations for the design and construction of concrete sea structures. Fédération Internationale de la Précontrainte, Wexham Springs, Slough, 1974.

Recommended practice for segmental construction in prestressed concrete. *J. Prestressed Concr. Inst.*, 1975, Mar./Apr.

Rules for building and classing steel vessels. American Bureau of Shipping, New York, 1975.

SOMERVILLE G. and TAYLOR H. P. *Concrete in the oceans: Reports 1 and 2*. Ship and Marine Technology Requirements Board, Department of Industry, London, 1974.

Story of the Selma, Expanded Shale, Clay and Slate Institute, Washington, DC, 1960.

Concrete afloat. Thomas Telford Ltd, London, 1977, 71–81

Nineteen pretensioned prestressed concrete barges constructed in the Philippines in 1964, 1965 and 1966 are performing extremely well since their initial service 10–12 years ago. Both dry and liquid cargo prestressed concrete hulls have been used for heavy ocean-going service and have been exposed to corrosive climate and severe damage environment in the Vietnam War. Repairs to concrete hulls after major collisions or hostile explosive activity are simple to perform and require minimal 'down time'. Nearly all repairs are done while the vessel is afloat, therefore eliminating the need for dry-docking time and expenses. The durable characteristics of these concrete hulls virtually eliminate dry-docking requirements for maintenance. Operational experience for over a decade has proven that these concrete hulls require no additional fuel for towing over conventional steel hulls of similar dead-weight capacities. The paper discusses the basic construction and operational characteristics of the world's first major barge fleet constructed of pretensioned precast concrete.

Operational experience with prestressed concrete barges

P. N. SARE*
A. A. YEE†

In the days of the early sailing ships, concrete was used as ballast or deck and shell covering in the iron and steel vessels. This experience proved the durable characteristics of concrete and thus many shipbuilders became interested in using concrete as a material for the basic construction of their vessels. In World War I the scarcity of steel encouraged the construction of large ships with reinforced concrete; about 85 000 tons of seagoing shipping was built during this period. During World War II, when the shortage of steel was again critical, over 200 concrete vessels of various tonnages and types were constructed for ocean-going use in Europe and America.

The Luzon Stevedoring Corporation (LUSTEVECO) has, among its fleet of 400 steel barges, four 2000 dwt reinforced concrete barges built during World War II. These barges had a light-ship weight of 2700 tons with dimensions of 269 ft × 49 ft × 17 ft. The hulls of these barges are 12 in. thick, thus resulting in a high light-ship weight and a low dead weight to displacement ratio. These barges had a light draught of 8 ft and a fully loaded draught of 14 ft and were unwieldy to tow. For these reasons, they are utilized as floating warehouses for storage of general cargoes such as equipment or machinery on deck, or molasses, cement and other general cargo in the tanks below the deck. Some of these concrete vessels were used as floating repair and machine shops.

* Vice President in charge of engineering, Luzon Stevedoring Corporation, Manila.
† President, Alfred A. Yee and Associates, Inc., Honolulu.

Because of the durability of these World War II barges, LUSTEVECO engaged the services of Alfred A. Yee and Associates, Inc. to explore the potential of constructing prestressed concrete hulls that would meet the requirements for light-ship weight and draught which would allow higher dead weight to displacement ratios within the limits of the navigational depths and bridge clearances in the Pasig River and other waterways in the Philippines. The final hull structural design arrived at from this research involved framing with a series of transverse structural steel frames supporting a 4½ in. thick pretensioned prestressed concrete outer shell. Compartmentation and bulkheading were framed with conventional steel plating and structural shape stiffeners. The 4½ in. thick concrete shell was prestressed longitudinally to develop the overall hull ben-

Table 1. Prestressed concrete barges built by LUSTEVECO, 1964–66

Number of units	Dead-weight tonnage and type	General dimensions
10	700 dwt, general cargo	110 ft × 47 ft 1 in. × 12 ft 1 in.
6	2000 dwt, general cargo	198 ft × 56 ft 8 in. × 12 ft 6 in.
3	2000 dwt, bulk petroleum products	198 ft × 56 ft 8 in. × 12 ft 6 in.

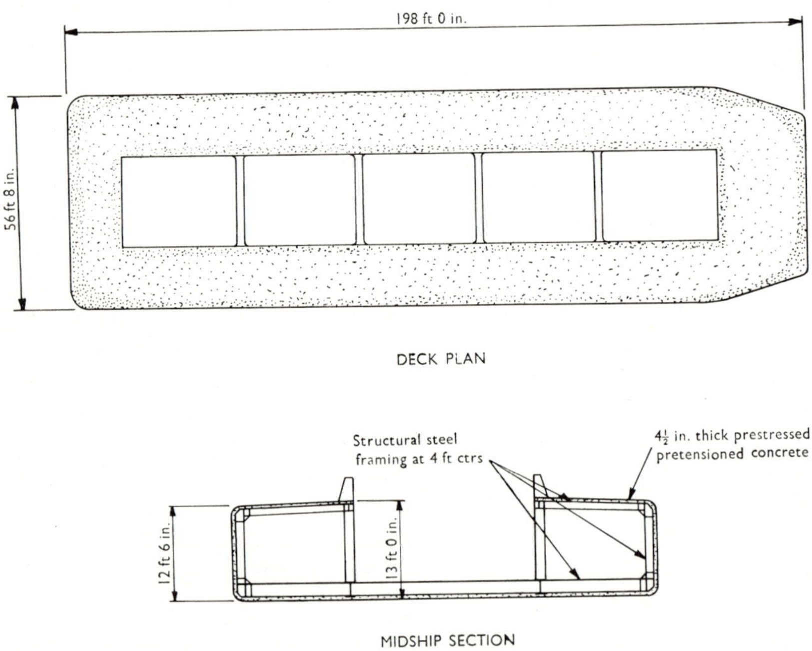

Fig. 1. Plan and section of 2000 ton general cargo barge

ding moments and local bending moments, and this shell was attached to the structural steel framing by means of welded shear lugs which were installed on the steel framing prior to the concrete shell pour. Both transverse and longitudinal conventional unprestressed reinforcing steel and wire mesh were added to the concrete shell for additional local and bending and impact resistance.

From 1964 to 1966 LUSTEVECO built, to American Bureau of Shipping class rules for ocean-going vessels, 19 prestressed concrete barges with particulars as shown in Table 1. Fig. 1 shows the basic dimensions and construction of the 2000 ton general cargo prestressed concrete barge.

Besides durability, the following factors heavily influenced LUSTEVECO's decision to pursue the building programme:

(*a*) low capital investment on construction facilities;
(*b*) ready availability of materials at low cost;
(*c*) low degree of skilled labour required, hence lower costs of labour;
(*d*) ease of repairs and minimal maintenance;
(*e*) minimal 'off-hire' or 'down-time' period due to ease of repairs;
(*f*) fire and explosion resistance.

TRADING HISTORY

Since their commissioning, these barges have seen service not only in the Philippines but in other areas of South-East Asia. In Vietnam, they were preferred by the military over steel barges for the carriage of ammunition and explosives because of the non-sparking characteristics of concrete. The 2000 dwt dry cargo units were very active in the transport of rice to Cambodia through the Mekong River; in addition, all structural steel highway bridge sections fabricated by RMK-BRJ at Poro, Philippines were loaded on these barges for transport to Vietnam—a 1200 mile ocean voyage which took 6–7 days. The three 2000 dwt tanker barges in service have proven themselves operationally superior in the transport of refined petroleum products. While the very corrosive petroleum products have wreaked havoc on the steel barges, which after a period of five years required extensive plate renewals, the concrete barges withstood this exposure without significant damage. The smaller 700 dwt units are extensively and successfully used for moving bulk industrial salts and fertilizers, which are both equally corrosive cargoes.

After 10–12 years at sea, steel barges generally look rather worn out as evidenced by numerous dents and corrugations and variously sized inserts unmistakably showing repairs of cracks and indentations. But not so with the prestressed units. The only visible sign of damage due to improper handling or poor seamanship is the occasional exposure of reinforcing bars by repeated abrasions of concrete. When this happens, a quick cement patch is applied and, after application of a coat of paint, the barge becomes as good as new.

CONSTRUCTION

Building costs are given in Tables 2 and 3. In the Philippines, the new building cost per dead weight is considerably higher for steel than for prestressed concrete and this is principally attributed to lower cost of material and labour. Also, it is more economical to build bigger concrete barges than smaller sizes. Furthermore, the length to depth ratio of 16 is maximum for ocean service and the breadth to depth ratio is rather high. Due to limitations imposed by local operating conditions, a maximum air draught of 12 ft and maximum water draught of 10 ft had to be maintained for the vessels to be able to go under the bridges at the Pasig River in Manila. Without these constraints, for the same length, the depth of the barge can be increased, thus increasing its cargo capacity and improving the dead weight to full load displacement ratio.

OPERATIONAL ASPECTS

Operations

Weight comparison between prestressed concrete and steel barges. Table 4 shows that for the same capacity the concrete barge has a 19–24% lower dead weight to full load displacement ratio; i.e., steel has a better cargo utilization factor than concrete. This is due mainly to steel being

Table 2. Building costs in the Philippines in 1976, US$

Dead-weight tonnage and type	Building cost, US$	
	Prestressed concrete	Steel
2000 dwt, dry cargo	180 000	353 000
2000 dwt, POL	210 000	420 000
700 dwt, dry cargo	80 000	120 000

Table 3. Building costs in the Philippines in 1976, US$/dwt

Dead-weight tonnage and type	Building cost, US$/dwt	
	Prestressed concrete	Steel
2000 dwt, dry cargo	90·00	176·50
2000 dwt, POL	105·00	210·00
700 dwt, dry cargo	114·30	171·45
Average	103·10	185·98

lighter than prestressed concrete. However, considerable gains have been made in reducing the weight. It is felt that with the use of lightweight aggregate, a decrease in light-ship weight by 10% can be attained. This would improve the dead weight to displacement ratio of the prestressed concrete barges considerably. Nevertheless, when high volume, low density cargo is to be transported, the prestressed concrete barges can be used with advantage over steel vessels due to the larger storage area of the concrete hulls.

Towing speed. Table 5 shows that by utilizing the same tug and maintaining the same fuel consumption, there is a decrease of towing speed of only 6–8% for prestressed hulls compared to steel barges. This is due to the smoother hull surface of concrete barges, better bow lines and the absence of skegs which increase hull resistance by at least 10%. Considering the relatively short distance between ports in the Philippines, which on average takes 36 towing hours, this reduction in speed means a delay of two to three hours which is easily offset by the fact that, unlike steel barges, prestressed concrete barges do not require ballasting and deballasting prior to being towed empty—an operation taking a total of five hours—thus a net saving of two hours can be realized by using prestressed concrete vessels.

Behaviour at sea. Steel barges sit high when operating light and therefore must be ballasted down in order to put the skegs in action to prevent yawing in the tow. The concrete barges sit deeper in the water due to their heavier light-ship weight, and therefore towing light incurs less yawing. In the longer prestressed concrete barges such as the 2000 dwt capacity types, it has been found that because of the added drag there is no need for skegs or drogs during the tow. The tugboat captains at LUSTEVECO have repeatedly preferred towing prestressed concrete barges because of their greater stability, less yawing, and less slamming.

These barges have weathered the worst ocean weather in the Philippines, specifically during the north-east and the more severe south-west monsoons, without appreciable damage. Compared to steel barges, they are less prone to yaw, vibrate, pound and roll under adverse weather conditions because of their greater weight.

Effect on cargo. Because concrete is not subjected to the corrosive effect of refined petroleum products, contamination of cargo from rust is kept minimal. This is very important to the oil industry where rigid tests of cargo samples are conducted prior to cargo discharge. We have had no instance where petroleum products were rejected by consignees because the cargoes were found, after laboratory test, to be off specifications. Where rust reaction takes place, foreign bodies and colour change may

Table 4. Comparison of weights (long tons) of prestressed concrete and steel barges

	2000 dwt, dry cargo		2000 dwt, POL		700 dwt, dry cargo	
	Prestressed concrete	Steel	Prestressed concrete	Steel	Prestressed concrete	Steel
Hull Deck hardware, fittings Equipment	930 10 5	377 10 5	970 10 5	405 10 5	447 7 3	160 7 3
Light displacement	945	392	985	420	457	170
Stores Cargo	— 2000	— 2000	— 2000	— 2000	— 700	— 700
Total dead weight	2000	2000	2000	2000	700	700
Full load displacement	2945	2392	2985	2420	1157	870
Percentage dead weight to full load displacement	68%	84%	67%	83%	61%	80%
	(Concrete 19% lower)		(Concrete 19% lower)		(Concrete 24% lower)	

Table 5. Towing data provided by LUSTEVECO's Marine Division Manager based on daily operation reports submitted by tug masters: towing tug, one coastal tug, 1200 bhp; fuel consumption, 47 gal./h diesel oil; condition of sea, slight to moderate

Tow	Average towing speed, knots
1 × 2000 dwt prestressed concrete barge (L-1960)	5·3
1 × 2000 dwt steel barge (L-1304)	5·6
3 × 700 dwt prestressed concrete barges (tandem tow)	3·7
3 × 700 dwt steel barges (tandem tow)	4·0

occur in the petroleum products and this can be harmful from the sales point of view or to the machinery which utilizes the product. Generally, should such cargo contamination occur, LUSTEVECO must bear the cost of pumping back the petroleum products to the refineries for reprocessing.

Concrete, having a lower thermal conductivity, is less susceptible to sweating inside the holds than steel barges. In the transport of bulk corn, copra pellets, bagged sugar and fertilizers, claims for cargo damage due to sweating are minimal in the case of prestressed concrete barges in comparison with those incurred when steel hulls are used.

Repairs and maintenance

Repair experience. The majority of damage to prestressed barges is due to direct contact with towing tugs and barges or when coming alongside piers. These damages are sustained at the side and their repairs involve the breaking up of about 3–4 ft^2 of concrete to expose all reinforcing steel elements, roughing up the parent surface, applying bonding compound and pouring concrete on the damaged area.

The most severe damage is that due to grounding and explosion. An illustration is the extensive repair done on one of LUSTEVECO's 2000 dwt dry cargo barges, L-1960 (Fig. 2). The barge was transiting the Mekong River in March 1973, fully loaded with rice intended for Phnom Penh, when she hit a mine at the starboard side of cargo hold 3. As a result of this explosion, three frames were bent inward by as much as 36 in., reinforcements snapped and concrete caved in on an area measuring 16 ft × 12 ft.

It was decided to conduct temporary repairs to enable the barge to be towed safely to the Philippines for permanent repairs. After beaching and unloading the cargo, all debris was cleared and a $\frac{1}{2}$ in. × 20 ft × 12 ft steel plate, stiffened by $\frac{3}{8}$ in. × 4 in. × 3 in. angles spaced 24 in. on centres, was placed and secured over the damaged area. One-inch bolts were then welded to the plate and secured on the deformed frames. Clay was applied along the perimeter of the covered hole inside and outside, thence grouted with concrete using high early strength mixture. After three days

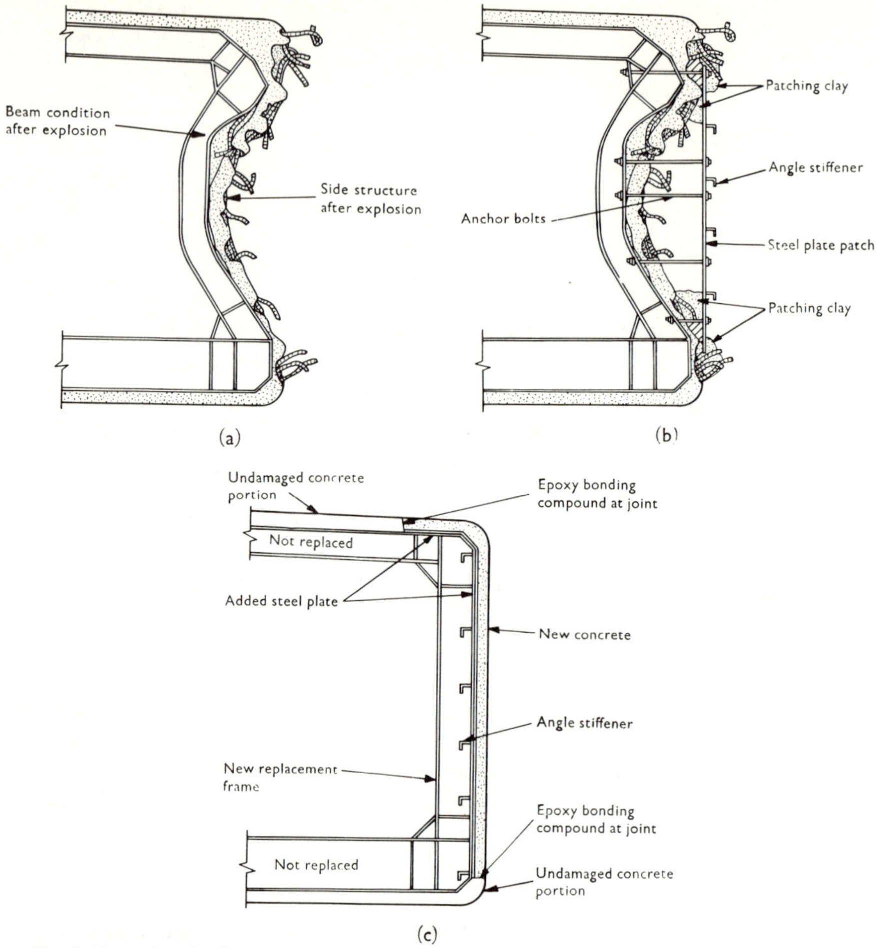

Fig. 2. Barge involved in mine explosion incident: (a) condition after explosion; (b) detail of temporary repair; (c) detail of repaired section

the barge was ready for towing back to the shipyard for permanent repairs.

On arrival at the shipyard, permanent repairs were conducted at the dry dock. The bent frames were cut and the loose concrete portion removed. The three frames were replaced with new identical structures, and $\frac{1}{2}$ in. thick plate stiffened with $\frac{3}{8}$ in. × 4 in. × 3 in. angles spaced 24 in. apart was welded on the frames, practically sealing the whole damaged area.

The No. 4 steel bar reinforcements were aligned and the cut sections were supplemented with new ones overlapping 30 in. Steel strands were likewise realigned. Broken sections were provided with new strands lapped 30 in. at both ends.

Table 6. Cost of repair of L-1960 (Philippines cost in 1976)

	Cost, US$
Material	2133
Labour	853
Dry-docking	1100
Others	295
Total	4381

Table 7. Estimated costs for repair of a steel barge in ten days (for comparison with Table 6)

	Cost, US$
6 tons steel at US$ 1400/ton	8400
Dry-docking	1100
Others	300
Total	9800

Side forms were then prepared and the exposed concrete surface cleaned of dust and debris and roughened. Sika's Colma bonding compound was applied and a high strength concrete was poured and vibrated. Forms were removed after seven days and the barge refloated and recommissioned on the tenth day.

The cost of repair is shown in Table 6. Estimated costs to complete similar repairs on a steel barge in ten days are given in Table 7.

A significant consideration is the repair of barges carrying low flash point petroleum products. An oil tanker steel barge which sustained a crack on its side shell plating would have to be cleaned and gas-freed before cutting and welding could proceed. The cleaning and gas-freeing might involve not only the cargo tank affected but the adjacent tanks as well. On a 2000 dwt steel tanker barge, this process takes eight days because gases are generally pocketed underneath rust scales, necessitating continuous butterworthing. If the same damage occurred in a prestressed concrete barge, no extensive cleaning and gas-freeing would be necessary except general washing to clear the fumes and providing sufficient ventilation inside the cargo tanks, which takes at the most two days. Considering that the daily charter of this size of barge is US$295, the saving of six days means an additional revenue to the company of US$1770.

Maintenance experience. The Philippine Coast Guard allows these prestressed concrete barges to operate without the required annual dry-docking provided that the afloat survey shows that the internals, bulkheads, fittings and the shell are sound and watertight. All the 19 prestressed concrete barges have never been dry-docked for annual survey since they were commissioned. Periodic afloat maintenance work such as renewal of fenders, replacement of gaskets, patching hairline cracks and painting are conducted while the barges are in service.

For this work, LUSTEVECO's records show that from 1974 to 1975 they spent an average of US$73·45 per barge per year. The total down time per barge per year for the above-mentioned afloat maintenance work is six days. On the other hand, the average down time due to repairs, per barge per year, of the 400 steel barges is 24 days and the average repair/maintenance cost is US $8167.

For vessels carrying corrosive cargo such as petroleum products or other low flash point products, experience has shown that a five year old steel-hulled vessel would have to have approximately 50% of its steel plating and 40% of its steel scantling replaced. For the prestressed concrete vessels, repairs and maintenance requirements have been only minimal. For vessels carrying general dry cargo, our experience has shown for a twelve year old steel-hulled vessel at least 50% of its steel plating and 30% of its scantlings would have to be replaced. Annual drydocking for a vessel of this condition would result in at least 35 days down time. The prestressed concrete barges have never been dry-docked for maintenance. In only a few instances for large major repairs has drydocking been required for the prestressed concrete vessels. Most of the repairs to the prestressed concrete vessels could be performed by ballasting up the hull while afloat. Maintenance requirements for prestressed concrete hulls carrying both liquid and dry cargoes have been minimal during the past 10–12 years.

GENERAL OBSERVATIONS

On the debit side, the following observations are submitted.

Hairline cracks on deck appear after about four years' service. In some instances these cracks admit sea water to the steel reinforcements and rusting occurs. This situation can be remedied by early detection of cracks, washing with fresh water, drying and painting the deck immediately with a coat of coal tar for waterproofing.

Damages more often are inflicted by collision with tugboats or other barges. Since the concrete shell cannot stand impact as well as steel, sufficient fenders should be installed at the sides.

CONCLUSION

It is not the intention of this paper to suggest that prestressed barges can totally replace steel barges. While it is true that prestressed barges have low initial building costs, comparable towing speed, better seakeeping characteristics, lower maintenance and repair cost, lower down time and consequently more revenue-generating days, their suitability will depend on the type of service, hydrographical condition and the cargoes to be transported. Whenever cargo utilization factor is not a primary consideration, such as for floating docks, dredges, floating cranes and oil drilling rigs, prestressed concrete would unquestionably be an ideal material to use. Furthermore, prestressed concrete barges will be very suitable for low density and high volume cargoes such as wood chips, coke and the like. Additionally, they are desirable in the transport of highly volatile and corrosive petroleum products. Where port congestion is prevalent, as in the Middle East, prestressed concrete barges could be economically built to serve as floating warehouses, processing plants or factories. However, in areas where draught is an important consideration or where the nature of the trade requires frequent shifting of cargo with high densities, steel barges would definitely be a more suitable equipment to use.

With this breakthrough in the construction and operation of prestressed concrete barges, and as better technology develops and man continuously seeks better ways of improving water transportation, the use of concrete as a shipbuilding material will continue to be an exciting challenge.

Concrete afloat. Thomas Telford Ltd, London, 1977, 83–96

The paper briefly describes the work that has been carried out since 1972 on a new concept for the transportation of liquid natural gas (LNG), using a self-propelled carrier with a reinforced concrete hull and insulated cargo tanks formed within the hull. The author concentrates on the economic advantages of the concrete vessel when compared with the conventional steel LNG carriers.

LNG concrete ship economics

A. E. STANFORD, BMS, SNAME(LM), FIMarE, FRINA*

DYTAM Tanker GmbH, a joint venture company of Tampimex Tankers Ltd (UK) and Dyckerhoff and Widmann (Germany), has been actively involved with the detailed investigation[1] of concrete as an alternative material for the construction of vessels for the marine transportation of liquid methane (liquefied natural gas (LNG)).

The choice of post-tensioned and reinforced concrete as a building material was based on economic, technical and ecological considerations: the first because of the soaring cost of conventional steel carriers, the second because of the highly desirable behaviour of concrete when subjected to cryogenic temperature conditions and the third because in events such as groundings, fire and collision, the concrete carrier offers greater protection to the environment than a steel vessel.

The vessel is of monolithic, concrete construction throughout the entire hull. The thick shell of the hull consists of high quality concrete and is reinforced longitudinally and transversely by stressed and non-stressed steel rods. The cargo tanks are integral with the hull structure and the cargo tank transverse bulkheads are also made of concrete. The vessel has a single hull. The hull or tank internal is separated from the cargo by insulation in which are incorporated primary and secondary barriers.

The conventional steel LNG carriers are of the double-hull design. The

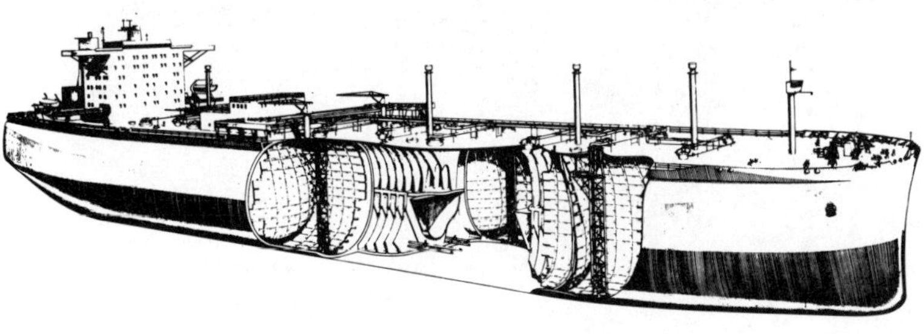

Fig. 1. Cut-away impression of 128 000 m³ concrete LNG carrier

* President, DYTAM Marine Inc., New York, NY.

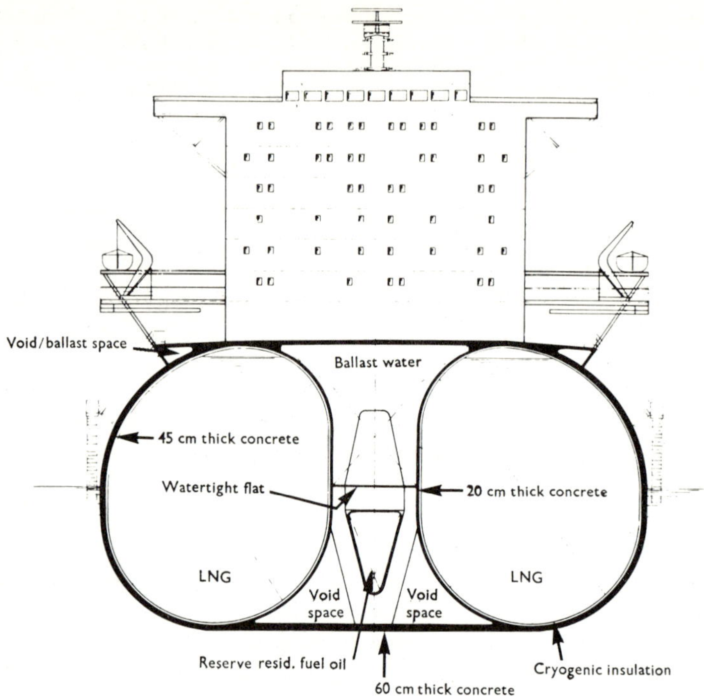

Fig. 2

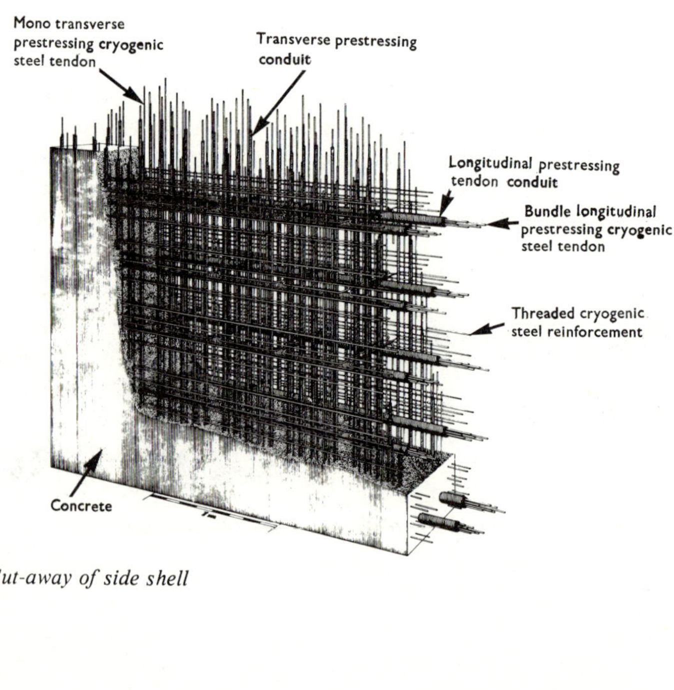

Fig. 3. Cut-away of side shell

outer hull is constructed of non-cryogenic steel. The external hull and ballast tank areas are generally coated with expensive corrosion-resistant paints and/or cathodic protective systems. To avoid marine growth the normal protection system for the bottom hull area is to apply an anti-foulant coating on top of the costly corrosion-protection paint application.

During the life of the steel carrier it is necessary to continuously repair and/or totally renew these coated areas. The total surface area is quite extensive; depending on the quality of the protective corrosion-resisting systems employed during the construction of the vessel, high steel replacement and coating restoration costs may evolve. The concrete ship requires 'no' protective coating throughout the entire external hull and all ballast tank areas, thus offering very attractive economic advantages to an owner for the life of the ship.

THE SHIP

Description

The monolithic concrete hull, as shown in Fig. 1, consisting of a high quality, dense concrete (Fig. 2), is reinforced longitudinally and transversely by stressed and non-stressed steel rods (Fig. 3).

Figure 4 shows the cargo tanks as an integral part of the hull structure, and the tank transverse bulkheads are also made of concrete. This vessel,

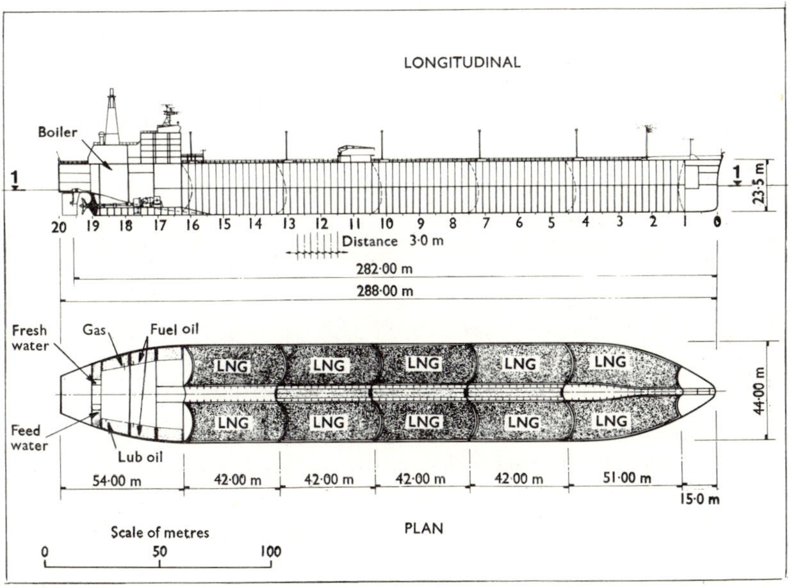

Fig. 4

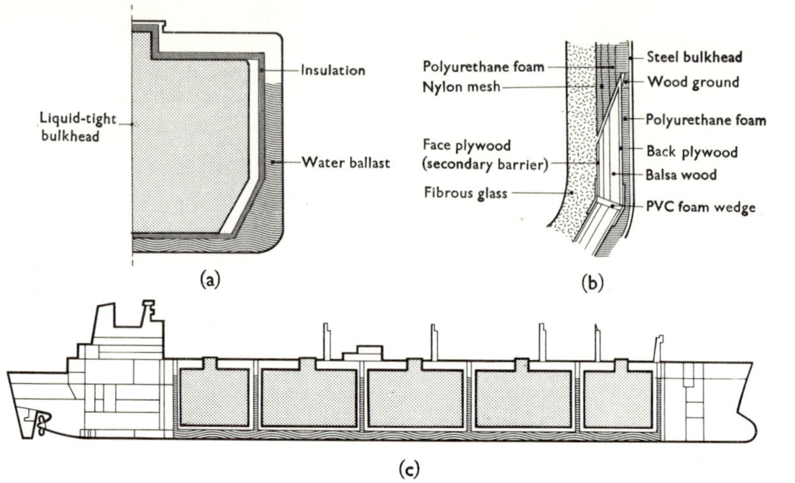

Fig. 5. Conch free-standing tank system (ballast tank area approx. 150 000 m² to 260 000 m²): (a) section; (b) detail of tank insulation; (c) profile (reproduced by permission of Fairplay Publications Ltd[8])

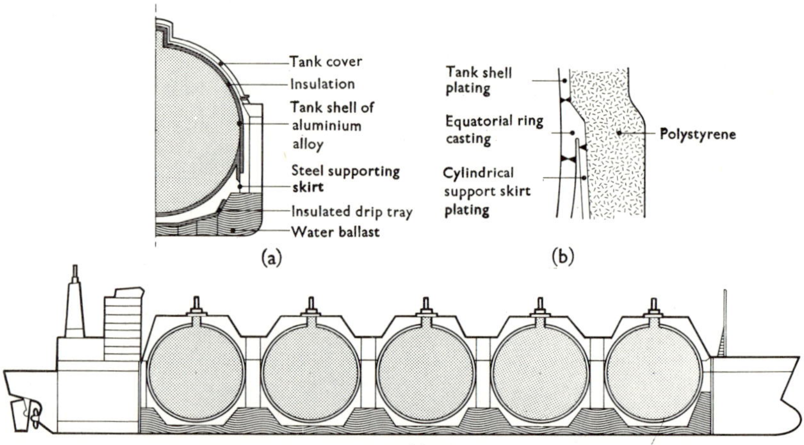

Fig. 6. Kvaerner–Moss free-standing spherical tank system (total ballast tank area approx. 180 000 m²): (a) section; (b) detail of tank support-skirt connection; (c) profile (reproduced by permission of Fairplay Publications Ltd[8])

unlike the conventional steel LNG carriers (as shown in Figs 5–7) is a single-hull design. The insulation system (Fig. 8) separates the cargo from the hull with its incorporated primary and secondary barriers and a space filled with inert gas. This insulation is of the load-bearing and impact-resistance type.

The overall dimensions and capacities of the vessel are shown in Table 1.

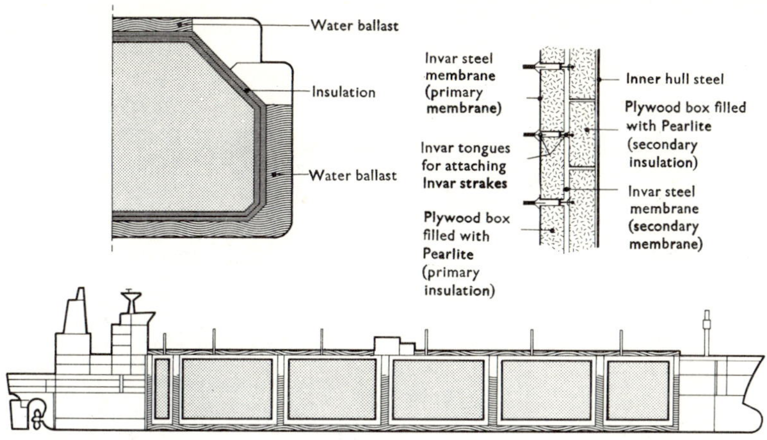

Fig. 7. Gaz Transport membrane tank system (total ballast tank area 215 000 m²): (a) section; (b) detail of membrane and insulation; (c) profile (reproduced by permission of Fairplay Publications Ltd[8])

Vessel strength

Of primary concern to Lloyd's Register, as the classification society, was the matter of longitudinal and transverse strength. Final selection of the arc form takes advantage of both the compressive strength and stability characteristics of concrete. This form lends itself to reduction of transverse bending stresses by utilization of membrane responses. With the use of steels with cryogenic temperature resistance for both post-tensioning and normal reinforcement, this leads to a very efficient structural solution.[2–4]

The concrete mixture to be employed is a dense, pumpable type, with a low water to cement ratio (max 0·45), a maximum aggregate size of 6 mm and a density of maximum 2·0 kg/cm^3 (wet). Concrete cover of reinforcement material is externally 3 cm, internally 2 cm.

All normal reinforcing and post-tensioning steel is of the cryogenic, high carbon type, having an ultimate strength of 15 000 kg/cm^2, a yield of 13 500 kg/cm^2 and an elastic limit of 12 000 kg/cm^2. Maximum tensile stress in the post-tensioning steel does not exceed about two thirds of the yield strength. In the normal reinforcement steel it is approximately 10% of yield. Both classes of steel are from Friedrich Krupp Huttenwerke, AG, type St 135/150.

Normal reinforcing rods are of screw thread type deformation, permitting use of threaded couplers, if desired. Post-tensioning is by seven-strand bundles in steel, externally ribbed conduits. Following post-tensioning the bundles are pressure-grouted within the conduits, thereby becoming integral with the hull.

The vessel has received full conceptual and detail approval of the en-

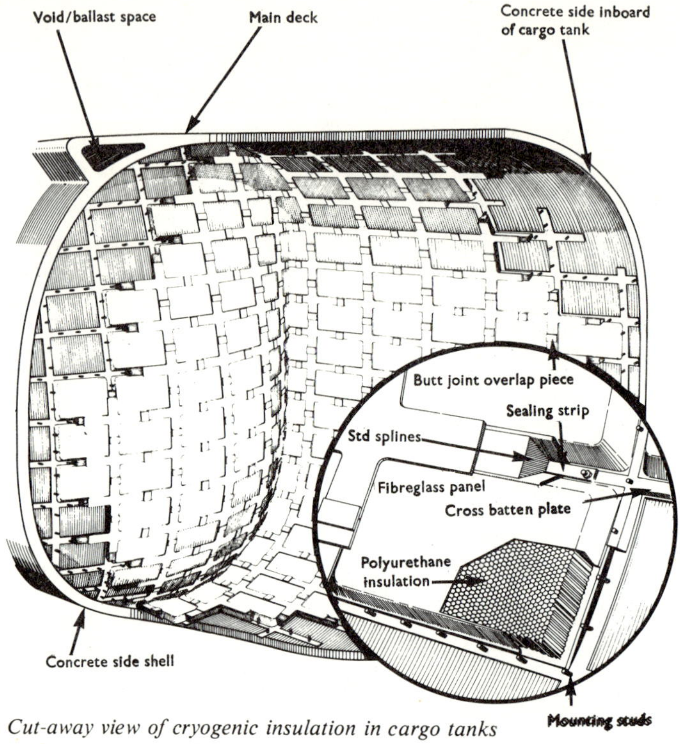

Fig. 8. Cut-away view of cryogenic insulation in cargo tanks

Table 1. Design data

Length overall	290 m
Length between perpendiculars	282 m
Beam (extreme)	44 m
Depth of hold (at centre line)	23·5 m
Designed draught (keel), summer load line	11·6 m
Cargo dead weight at design draught	56 250 t
Gross registered tons (grt) approx.	61 000
Cubic capacity required for LNG (56 250 t) at 0·45 cargo specific gravity and 11·6 m draught	126 875 m^3
Residual fuel oil tank capacity (98%) (includes approx. 2000 t reserve)	4000 t
Shaft horse power (shp) (metric max. continuous)	40 000*
Cruising period on residual fuel (after service tanks only) at 40 000 shp, approx.	10 days
Speed (loaded—trial)	19·5 knots plus†

* This shaft horse power specification is subject to final model tank testing and owner's/operator's requirements as to desired speeds relative to possible fuel economies.
† This loaded trial speed specification at 40 000 shp is based on predictions by The Hamburgische Schiffsbauversuchsanstalt, GmbH (HSVA), Hamburg.

tire hull structure by Lloyd's, and single-hull concept approval by the US Coast Guard, and complies with all present IMCO requirements, where applicable to a concrete carrier.

OPERATIONS

Operational advantages of the concrete vessel over the conventional steel LNG carrier are many. These include, but are not limited to, the following.

(*a*) The concrete vessel has a longer, basically maintenance-free (by comparison) hull life, exceeding 25 years. Some 56 year old hulls are still afloat and serving.

(*b*) The need for painting or coating is virtually eliminated, other than for steel superstructure, deck machinery and fittings, certain deck piping and hangers, bow and forefoot wrapper plates, anchor striking plates, rudder and skeg extremity boot. Painting or coating of deck, side shell and underwater body is purely at an owner's wish; not a necessity, as for a steel vessel.

(*c*) Historically, concrete vessels have proved to be exceptionally sea-kindly. Their response to the magnitude and frequency of varied sea spectra has been shown to be less violent and immediate than the equivalent steel vessel.

(*d*) A very noticeable characteristic of the concrete ship is its lack of vibration and its sound-muffling effect.

(*e*) The very elastic nature of the steel reinforcement and prestressing, with its ability to hold fractured concrete in place, tends substantially to reduce the effects of internal explosion, collision or grounding, thus cutting down on the loss of cargo should such accidents take place. In this connection, possible egress of product is further substantially reduced by this carrier's smaller tank size.

(*f*) Repairs to a concrete vessel are substantially less costly and more easily and rapidly effected, and require less dry dock time than an equivalent steel vessel for similar damage. Both reinforcement and prestressing steel need be replaced only in the immediate area of the damage. No loss of reinforcement strength or prestressing compression occurs outside the immediately damaged area.

(*g*) If a concrete vessel takes the ground lightly, or touches bottom when crossing a bar or shoal, scarping and scoring effects are of little consequence until the next scheduled dry docking. With the steel vessel, paint or coating systems are disturbed and wastage of steel commences.

(*h*) With the DYTAM single-hull concrete LNG carrier a very desirable feature as compared to competitive and comparative steel carriers is the low profile. This reduces leeward set at sea and lessens docking hazards in high winds.

(*i*) The concrete flush deck arrangement permits a simple, accessible, easily maintained deck piping layout. Deck wiring for both power and lighting requirements can be kept to a minimum.

(*j*) As the ballast tank spaces are of concrete, the problem of steel corrosion within these spaces is eliminated. This also applies to the side shell, inner bottom tanks and bilges within the machinery spaces. The same holds true for fore and after peak tanks, stores and steering gear spaces.

Operational days per year

The annual utilization of a ship varies appreciably from year to year. However, it is accepted that the conventional gas carrier will average approximately 340 days over the ship's life.[5,6]

As the machinery, deck equipment, cargo system, navigational equipment etc. are basically the same for both steel and concrete hull ships, no repair benefit can be claimed for the concrete ship in these areas. The advantage from this standpoint would be due exclusively to the hull material.

Figures 5–7 show the typical ballast tank arrangements for conven-

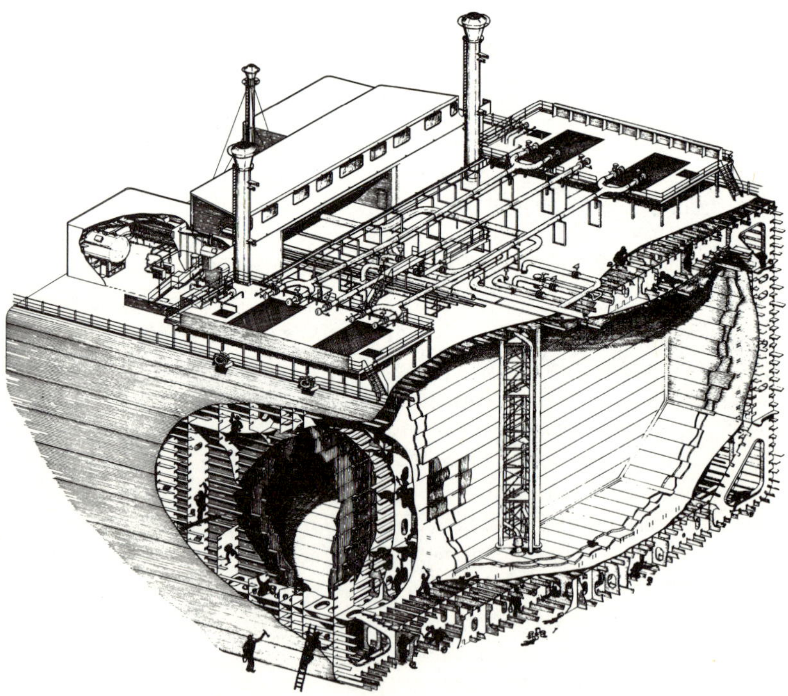

Fig. 9. Cut-away view of Gaz Transport membrane tank systems

tional steel LNG carriers. Fig. 9 illustrates in more detail the complicated steel structure within the ballast tank areas. One can appreciate the difficulty that arises when trying to sandblast and recoat the corroded areas within these confined spaces. Steel replacement repair time and cost will be very substantial when carried out in the shipyard.

For the conventional ship over its 20 year life, two complete renewals of the ballast tank coating are expected. This work, which is difficult to carry out because of the poor accessibility and lack of space, and because of the large area to be sandblasted and recoated (approximately 250 000 m^2), requires off-hire periods of at least 180 days over the ship's life.

It is estimated that on four occasions during the ship's life, periods of 14 days will be needed for extensive steel renewal, resulting in a total steel renewal time of 56 days.

Further, approximately every 14 months the steel ship will be dry-docked for bottom inspection, cleaning and renewal of the anti-fouling system. Deducting the four dockings required for steel renewal gives 13 dockings, each of five days' duration, and thus a total of 65 additional outages over the ship's life.

Therefore the total time over 20 years required on the steel ship's hull exclusively is 300 days—15 days per year more than for concrete carriers. We at DYTAM feel that a total of 13 days a year more for operation of the concrete carrier (i.e. 353 days per year) is a conservative estimate to include in our economic evaluation.

INSULATION

Because of the cryogenic properties of prestressed concrete, the safety requirements for the insulation system can be considerably less demanding than for a steel ship. In principle any acceptable insulation system presently available can be used. However, we have found that the Owens-Corning system (Fig. 8) is presently the most suitable for our vessel.

ECONOMICS

Using the computer program of Marine Service (Hamburg), an optimization study[7] was carried out comparing steel and concrete designs of approximately the same transportation capacity.

Three different trade routes were analysed, and it was found that the overall economic advantages of the concrete vessel for a 20 year period are similarly attractive for each. Therefore this paper concentrates on the Arzew (Algeria) to Lake Charles (US Gulf Coast) trade route.

Assuming the end of 1978 as a delivery date, the estimated selling price of a concrete LNG carrier is approximately 12·5% lower than the average for a steel carrier of similar capacity (Table 2).

Obviously, to appraise realistically such designs, the prospective owner will be interested in the full life cycle of the ship and, therefore, the economies are considered over a 20 year life.

Due to the excellent advantages of concrete from a corrosion-resistant standpoint, as described in the operational section, it is expected to operate for 353 days per year as compared with 340 days for the steel carrier. Additional to the benefit of more operational days per year, the difference in maintenance and repair cost between a concrete LNG carrier and a conventional steel ship is due to the preservation of the latter's ballast tanks, the underwater paint system and the general preservation of the remaining part of the steel hull (Tables 3 and 4).

Due to the unique hull form, we have been able to obtain a very competitive operating speed, as substantiated by Hamburgische Schiffsbau Versuchsanstalt GmbH (HSVA), Hamburg. For the purpose of this study, however, we have charged the concrete hull with approximately 4600 bhp more for the equivalent speed (19·5 knots) of the steel carrier. This again reflects a most conservative approach for our carrier.

For the specific trade Arzew to Lake Charles, Tables 2 and 5 reflect

Table 2. Cost comparison: trade Arzew (Algeria) to Lake Charles (US Gulf Coast) (optimized); one vessel; contract quantity 95 000 10^3 ft^3/day

	Years 1–12		Years 13–20	
	Concrete ship	Steel ship	Concrete ship	Steel ship
Service speed, knots	16·6	17·9	16·6	17·9
Round voyage time, days	27·3	25·5	27·3	25·5
Operation days per year	353	340	353	340
Voyages per year	12·9	13·3	12·9	13·3
Vessel capacity, m^3	128 200	124 300	128 200	124 300
Engine output, bhp	28 483	31 100	28 483	31 100
Ship price,* US$	104 000 000	119 000 000		
Maintenance and repair cost, US$/year	453 000	1 000 000	550 000	1 750 000
Insurance cost, US$/year	1 170 000	1 260 000	1 170 000	1 260 000
Operational cost,† US$/year	2 963 000	3 626 000	3 620 000	4 370 000
Bare-boat charter cost, US$/year	17 660 000	20 200 000		
Time charter cost, US$/year	20 642 000	23 823 000	3 620 000	4 370 000
Bunker fuel cost, US$/year	595 000	858 000	595 000	858 000
Boil-off fuel cost, US$/year	3 380 000	3 213 000	3 380 000	3 213 000
Port charges, US$/year	423 000	495 000	423 000	495 000
Transportation cost per cubic metre of liquid, US$	16·56	18·66	4·9	5·9
Transportation cost/10^3 ft^3, US$	0·76	0·85	0·226	0·27
Transportation cost/10^6 Btu, US$	0·67	0·76	0·20	0·24

* Final ship price based on delivery in 1978.
† Operational cost based on 1976.

Table 3. Cost for hull and ballast tank maintenance for steel ship only: years 1–12

	US$	US$
Ballast tanks: repair on coating system		
Years 1–7	798 000	
Year 8: sandblasting and renewal of complete coating system	2310 800	
Years 9–12	548 000	
Years 1–12, ballast tanks		3656 800
Hull		
Cleaning, repair and/or renewal of underwater paint systems and anti-fouling every 14 months: years 1–12	1413 000	
General hull maintenance, excluding accommodation: years 1–12	402 200	
Years 1–12, hull		1815 200
Steel renewals		
Year 8	550 000	
Year 12	550 000	
Years 1–12, steel		1100 000
Total cost for years 1–12		6572 000
Annual cost for years 1–12		547 666

Table 4. Cost for hull and ballast tank maintenance for steel ship only: years 13–20

	US$	US$
Ballast tanks: repair on coating system		
Years 13 and 14	496 000	
Year 15: sandblasting and renewal of complete coating system	3589 000	
Years 16–20	1185 000	
Years 13–20, ballast tanks		5270 000
Hull		
Cleaning, repair and/or renewal of underwater paint system (antifouling): years 13–20	1508 000	
General hull maintenance excluding accommodation	357 000	
Years 13–20, hull		1865 000
Steel renewals		
Year 15	780 000	
Year 18	1720 000	
Years 13–20		2500 000
Total cost for years 13–20		9635 000
Annual cost for years 13–20		1204 375

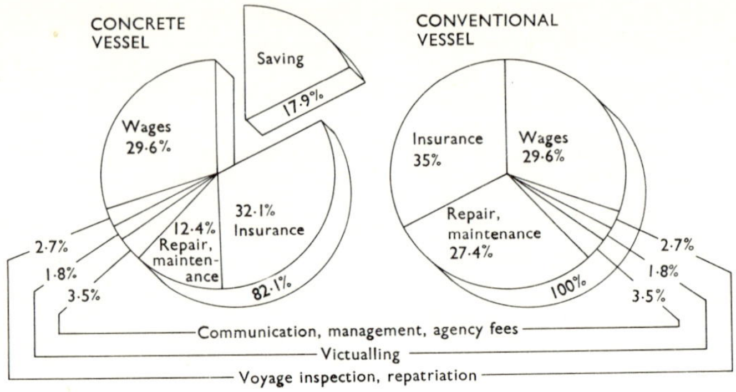

Fig. 10. Total operation costs

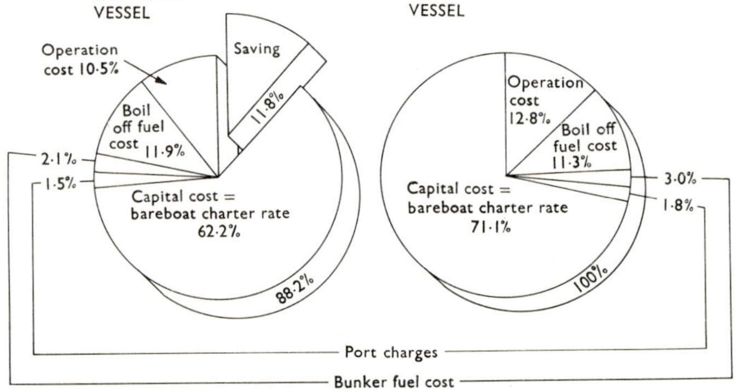

Fig. 11. Total annual cost

Table 5. Summary of approximate savings over 20 years for concrete ship compared to steel LNG carrier: trade Arzew to Lake Charles; optimization basis

	Years 1–12	Years 13–20
Ship price, US$'000	15 000	—
Maintenance and repair cost per year, US$ '000	547	1200
Insurance cost per year, US$ '000	90	90
Operations cost per year, US$ '000	663	750
Bare boat charter cost per year, US$'000	2540	—
Time charter cost per year, US$ '000	3180	750
Bunker fuel cost per year,* US$ '000	263	263
Boil-off fuel cost per year (loss),† US$ '000	(167)	(167)
Port charges per year, US$ '000	72	72
Transportation cost per cubic metre of liquid, US$	2·1	1·0
Transportation cost per 10^3 ft^3, US$	0·09	0·044
Transportation cost per 10^6Btu, US$	0·1	0·04

* Bunker fuel cost US$ 85/ton.

† Boil-off fuel cost US$ 1·45 per 10^3 ft^3.

the overall economics to a shipowner and eventual cost savings to the end user over the 20 year life of the vessel.

Figure 10 shows the percentage influences that the maintenance costs etc. have on the operating costs. When comparing the steel vessel to the concrete ship, there is approximately a 50% saving in repair costs for the concrete vessel. This amounts to over US$ 550 000 per year during the first twelve years and rises to US$ 1200 000 for the remaining eight years. This is based on using present-day costs of US$970/ton for steel renewal and US$ 11·60/m^2 for coating repairs with no credit being taken for the effect of escalation over the life cycle.

The economics in this case, which would yield a 17% return on investment, are based on 30% owner's equity requiring a 70% loan with an 8·5% interest rate and a repayment period of seven years.

Taking the figures shown in Table 2 and Fig. 11, the annual transportation cost when using a concrete vessel will equate to approximately US$ 0·09 per 10^3 ft^3 or 11·8% less than for the steel carrier. Even if one takes both vessels to have the same capital investment cost, by using the operational savings alone, the transportation cost advantage to the concrete ship would be US$ 0·016 per 10^3 ft^3 or over US$ 550 000 per year, with the operating costs based on 1976 prices without escalation.

FLOATING STRUCTURES

To supplement the work carried out on the LNG carrier, prior to accepting concrete as a realistic material for the transport of LNG, a comprehensive, historical and technical analysis was carried out by Mr William C. Carvill (MARADVICE), who is the naval architectural consultant for DYTAM. His findings, after he had travelled the world inspecting and researching concrete vessels that had been in service in some cases as long as 50 years, confirmed that there was minimum maintenance required for the concrete portion of the ships. Where steel was used, however, such as for hand rails, bollards, machinery bed-plates etc., these items needed considerable maintenance.

. In the cases where retired concrete vessels were used alongside steel hulls for the purpose of floating breakwaters, the concrete hulls had a non-comparative longer life, both from the strength and corrosion-resistance standpoint. Additionally, the marine growth on the steel hulls in the same local environment was considerably greater over a given period than that found on the concrete hull surface.

CONCLUSION

Reinforced concrete, because of its superior ability to resist corrosion alone, offers far more economic advantages than steel for a LNG carrier. At the same time, the owner has a ship which is a safer, more durable and ecologically superior vessel for the transportation of LNG.

REFERENCES

1. STANFORD A. E. *et al.* A concrete answer to LNG transportation. *Gastech 76: proceedings*. Gastech Exhibitions Ltd, Rickmansworth, 1977.
2. *Hull structure report No. 4*. Lloyd's Register of Shipping, London, 1974.
3. EVANS J. H. and BOUFOUNOS T. P. *The resistance of a reinforced concrete cylindrical shell to penetration by a knife edge load.* Report produced for DYTAM by MIT, 1975.
4. EVANS J. H. *Memorandum on the overall structural design of the post-tensioned, reinforced concrete, 125 000 m³ LNG carrier.* Report produced for DYTAM by MIT, 1975.
5. *Analysis of LNG marine transportation for the Maratime administration.* Booz Allen, Applied Roomch Inc., USA, 1973.
6. SOESAN J. M. and PANKS D. LNG tanker: terotechnology as applied to design maintenance and operation points on practical conclusion. *Gastech 75: proceedings.* Gastech Exhibitions Ltd, Rickmansworth, 1976.
7. *Economic feasibility study for Tampimex Tankers.* Marine Service GmbH, Hamburg, 1976.
8. CORKHILL M. *LNG carriers: the ships and their market.* Fairplay Publications Ltd, London, 1975.

BIBLIOGRAPHY

FINSTERWALDER U. Betonschiffe in Schalenbauweise. *Der Deutsche Ingenieur*, 1947.
How concrete ships have workd. *Literary Digest*, 3 April, 1920.

Concrete afloat. Thomas Telford Ltd, London, 1977, 97–104

The financial preliminary survey for a vessel should include investigation of various factors, including the reputation of the shipbuilder and his access to suppliers and credit, delivery dates, availability of special shipbuilding finance, the pattern of progress payments, fabrication off the site and the availability of government subsidies. The provision of shipping finance is governed by the Industry Act 1972, and the paper discusses the outline arrangements, management of the scheme, eligibility for grants, guaranteed loans, security, documents, procedure for the purchaser, legal considerations and expenses. As an alternative to the purchaser being the owner and user of the ship, the possibility of making a leasing arrangement is discussed.

Some aspects of financing concrete vessels

F. R. WELSH, MA, FBIM, FRSA*

From a historical point of view, the materials now used in shipbuilding are of relatively recent development. Wood was used in a variety of ways for thousands of years. The industrial revolution provided builders, designers and users with iron and then mild steel. Each has its advantages, but within 200 years of the first use of iron and the later development of steel, we are now looking at another material that has notable advantages. Historically, concrete has been used widely and with great success for docks, bridges and other structures which are submerged partially in water. We have also seen the development of large concrete structures for use in the North Sea oil and gas industry. Here, concrete has proved its ability to contend with extremes that weaken steel, and concrete using steel as a reinforcement is now an accepted material for use at sea. Concrete is also used to provide a protective cover and ballast for oil and gas pipelines because of its strength and resistance to marine damage.

FINANCIAL PRELIMINARY SURVEY

There has been much publicity on the problems of the tanker owners who borrowed large sums long term against the security of the ship so purchased, without arranging suitable charters. The failure to find cargoes meant that they could not make capital repayments or even service the debt, and this in turn created difficulties for the lenders. Security for a loan of this kind is not always necessary where the lender is relying on corporate credit, provided it is of high quality, instead of the ship as security. Failures of this kind, often created by too much money chasing too few tankers with long term charters, has naturally made the lenders very cautious. The present lenders are the ones that survived and they

* Director, Fraser Ansbacher Ltd, London.

survived by being conservative.

The financing proposition and justification for building a ship should, therefore, demonstrate profitable life for the ship, bound up either by a long term charter, which takes account of future inflation, or by corporate credit and earnings that will be sufficient to cover interest and capital repayments if the ship is without employment. The loan agreement will record the lender's ranking in the event of a liquidation.

The shipowner must therefore have covered in outline all aspects of the ship's future life with the charterer or customer. He will, no doubt, also have been in close co-operation with a shipbuilder in arriving at the specification and cost of building and running the ship. So far, the project will have cost relatively little, because the problems will have been explored only on paper; little labour, and only cheap materials will have been used.

The bank, or if it is a large ship or a fleet of ships, a syndicate of banks, will need to know much more about the background when working out a financial package. The most important items are

(*a*) the reputation of the shipbuilder;
(*b*) the access the shipbuilder has to suppliers, the credit he can obtain, and his ability to fund the timing differences between payment of a subcontractor and the progress payments made to the yard;
(*c*) delivery dates;
(*d*) availability of special shipbuilding finance;
(*e*) the pattern of progress payments; and
(*f*) fabrication off the site and the availability of government subsidies.

I should like to go through each of these in turn, starting, however, with the present problems relating to banking syndicates. There are at present outstanding some law suits in the USA between the members of a banking syndicate and the lead bank for the syndicate, concerning the loan that the syndicate made to a Greek shipping group. The problems of legal interpretation and banking practice are not a matter for discussion here, but suffice it to say that the law suits would have been less likely to occur if the shipping group had not been in financial difficulties. The suits, which relate to financial disclosure and the definition of a security under US law, have had the effect of making syndicate financing more difficult at present, and if the suits succeed this will cause any future syndicate to be more careful about the borrower and the terms of the loan.

The content of this paper is made in the context of UK agencies, institutions and shipbuilders which vary from those of other countries.

Reputation of the shipbuilder

The reputation of the shipbuilder is of great importance. Most of the problems discussed below can be affected by his actions. Success or failure in design, assembly, building and commissioning will have a considerable effect on the cash flow of the project. Late delivery has a bad effect on the economic justification of the ship because the earliest earnings after completion are discounted by a smaller factor when calculating the internal rate of return or net present value of the ship in the financial analysis. In the case of a concrete ship there may be many additional problems. First of all, who is a shipbuilder in this context?—one of the well known names?, or a firm of civil engineers?, or a combination of the two? It is interesting to note that none of the established 'steel shipbuilders' have arranged to speak at this conference in order to present themselves as suitable builders of concrete ships. Nevertheless, it is unwise to neglect the skills of the established shipbuilders when building a concrete ship. The assembly and fitting of the fuel systems, steering gear, fire-fighting equipment and electronics, and many other matters of detail are all important and are, of course, reviewed by a registration agency before it gives a certificate of seaworthiness. Shipbuilding depends on an infrastructure of subcontractors undertaking specialized services and making specialized products. These subcontractors must be known and their problems understood.

Access by the shipbuilder to suppliers and credit

The shipbuilder must have the confidence of his subcontractors if he is to be able to obtain normal, or better than normal credit from them and thus be able to fund his share of the building contract. The progress payments made to the bank will have been arranged beforehand.

Delivery dates

Reference has been made to the effect that late delivery has on a discounted cash flow calculation. Penalty clauses are essential to recoup the losses, but inevitably settlement takes a long time. Late delivery also usually indicates a cost over-run which may have a bad financial effect on the project and, in turn, jeopardize the whole contract. A very thorough analysis of the new building by the shipowner should be prepared, if necessary by consultants, if there is any doubt about the competence of the yard.

Availability of special shipbuilding finance

The availability of special shipbuilding finance is vital if the ship is to be competitive in world markets. Subsidized credits are available for new ships built in the UK provided that the owner and yard are prepared to abide by certain regulations. The UK tax system also allows shipowners attractive benefits, the main benefit of which can be obtained by leasing if the company is unable to absorb the tax allowances at the first opportunity.

Pattern of progress payments

The pattern of progress payments is an important part of the financing package, and should be considered at the outset.

Fabrication off the site and the availability of government subsidies

Many parts and sections of a ship may be constructed off the site of the shipyard. It is important to be sure that advantage is taken of all the subsidies that are available and that none are lost through construction or assembly at a site where subsidies are unavailable or are reduced.

INDUSTRY ACT 1972

The provision of shipping finance is governed by the Industry Act 1972, and in particular, by Section 10 of this act. The Department of Industry published a booklet on this section in 1974 called *Industry Act 1972 (Section 10): guarantees for the construction of ships or mobile offshore installations*.

The statutory provisions of Section 10 of the act enables the Secretary of State for Industry to guarantee, with the consent of the Treasury, the repayment of loans and related interest made to UK bodies for the purpose of financing construction in the UK of new ships and mobile offshore installations. The total liability in respect of loans under guarantee may not exceed £1000 million, but this amount may be increased to £1400 million, in certain circumstances.

A shipowner intending to build should make preliminary arrangements with his bankers for the loan. The banks, in conjunction with the Secretary of State for Industry, have arranged that, where a guarantee is provided by the Secretary of State, any loan for a ship will be at a fixed rate of interest (currently 7½%), payable half-yearly. The interest rate for loans for offshore structures will be at the current rate for buyer credits guaranteed by ECGD for exports other than ships.

The Secretary of State is represented by the Ship Mortgage Finance Company (SMFC) Ltd, which

(*a*) negotiates the terms of the loan and, if necessary, the building contract;
(*b*) assessess credit-worthiness;
(*c*) advises on security;
(*d*) considers the loan, security documents, charter party (if any), and insurance cover and relevant terms of the building contract;
(*e*) prepares the guarantee;
(*f*) administers the guarantee and advises on any action necessary to minimize loss.

An advisory committee principally staffed by the SMFC considers applications for guarantees and advises the Department.

Eligibility

There are strict rules governing the eligibility of grants, the objectives of which are to ensure that finance is only provided for a bona fide order from a business subject to UK legislation, and to provide employment in UK shipyards.

Firstly, there is a size and power limit: a ship other than a tug or offshore structure must be of at least 100 tons, whilst a tug must rate at least 500 brake horsepower.

Completion of orders previously started are eligible, but repair and refitting is ineligible under the section. In this case, an application for financial assistance under Section 7 or 8 of the 1972 act might be made. However, it is a cumbersome procedure, and at this time it appears to be uncertain whether an application would be accepted by the Department of Industry.

The act requires that any ship or structure must be built in the UK, excluding the Isle of Man and the Channel Islands. The purchaser must be resident in, or be a company incorporated in any part of the UK. Definitions are included in the act for ship and mobile installation; both terms are self-explanatory, but it is advisable to check in case of doubt, in order to be sure that the new building is covered by the provisions of the act.

Construction of a ship or structure must be by an approved shipyard. So far, it appears that no conventional yard has gone in for large-scale production of concrete ships. The building skills and methods are different. Consequently, we have seen the development of new yards for concrete structures for the North Sea rather than building in traditional shipyards.

Guaranteed loans

There are certain restrictions attached to the guaranteed loans. Most notably, they cover only 70% of the cost of a ship (with certain restrictions for fishing vessels), or 80% of the cost of an offshore structure.

Generally, the loans are repayable over seven years for a ship and five years for an offshore structure. For small loans, less than £1 million for a ship, and £250 000 for an offshore structure, the repayment period is usually reduced. However, if a buyer can prove that he could obtain longer credit from an overseas builder, credit terms may be lengthened.

Repayment is normally by equal half-yearly instalments.

During the building of the ship, advance payments made are covered by the guarantee. The amount and timing relate to the work completed by the builder. The purchaser is expected to put up the first 10% of the contract price before drawing on the loan, and to have paid his 30% of the contract price, in the case of a ship, by the time of delivery.

There are certain restrictions relating to the purchase of foreign goods and services in the completion of the ship or structure, which may result in a reduction in the amount of the loan, and so force financing the foreign components onto the purchaser.

Security

The security to be provided for the loan depends on the financial position of the purchaser. In general, a charge is taken on the asset whilst it is being built, and when it is completed a first mortgage is taken during the time that the loan on the asset is outstanding.

Additional security may be required where a purchaser is in a weak financial position; this may be taken in several ways.

Documents

The documents are standardized as far as possible, but each will take account of the idiosyncracies of a loan. The documents are a financial agreement between the bank and the purchaser, a deed of convenant, and an owner's agreement between the Secretary of State and the purchaser under which the purchaser undertakes to pay certain expenses incurred by SMFC in arranging the loan, and to provide the background data.

There is another guarantee between the Secretary of State and the bank providing the loan; though esential to the whole scheme, it is not of direct concern to the purchaser or builder.

Procedure for purchaser

It is necessary to follow the instructions provided, but in following them a considerable amount of background work must be completed before making an application. The official steps are as follows.

As soon as there is a firm prospect of an order being placed with a UK shipyard, both the SMFC and the bank should be advised. The SMFC will then provide an application form. An undertaking should be obtained from the bank that a loan will be made available conditional on the bank receiving a guarantee from the Secretary of State.

The procedure is then for the completed SMFC form, the bank's conditional undertaking and the builder's quotation to be sent to the SMFC. The SMFC will then come to a conclusion about the provision of a guarantee, if at all, and the terms of the loan.

The purchaser has the opportunity to accept or refuse the terms. Assuming they are accepted, a draft financial agreement is prepared for consideration by the purchaser and bankers. Agreement of the draft allows the owner's agreement to be completed, the guarantee to be issued by the Department, and the financial agreement to be completed. Advances may then be made to finance construction. On completion the mortgage is registered and the deed of covenant is completed between the bank and the owner.

Legal considerations and expenses

The purchaser is responsible for payment of all legal charges and expenses, including those incurred by SMFC and the Secretary of State, and stamp duty. An intending purchaser is also liable where expense is incurred but no loan is taken up.

There are various charges relating to the loans for which the purchaser is liable when it is taken up. They are one third per cent of the principal to be paid to the Department, and a commitment commission of one per cent of the principal.

The purchaser may be asked to pay a negotiation fee to the bank of one per mille on the maximum principal of the loan; this is only payable where the buyer has no significant account with a clearing or Scottish bank. In certain circumstances, a management fee of £500/year is charged if the loan is syndicated.

Where a company's management and control are outside the UK it is necessary to obtain consent to borrow under the terms of the Control of Borrowing Order 1958. It is also necessary for a company which is resident in the Scheduled Territories, but is controlled by a non-resident, to obtain permission to borrow under Section 30 of the Exchange Control Act 1947.

LEASING

So far, the purchaser has been regarded as the owner and user of a ship. This need not necessarily be the case. In circumstances where a purchaser is unable to absorb the benefits of free depreciation on new UK building at the first opportunity, it may make financial sense to enter a leasing arrangement with a bank or banking partnership. In this case a would-be purchaser arranges for a bank or partnership of banks to purchase the ship to be built to the user's specification. The purchaser enters a leasing contract with the banks which enables him to use the ship in just the same way as he would have done had he owned the equity of the ship. Instead of repaying a loan in semi-annual instalments plus interest, he pays a leasing charge, all of which is an allowed deduction for UK taxation, which reflects the implied payment by the banks of the instalments and interest on the loans which will include a mark-up to cover the banks' costs and a profit.

The advantage comes from the bank or partnership of banks' ability to absorb the maximum depreciation immediately against their taxable profits. The banks gain by being able to defer their tax liabilities, whilst the user of the ship pays a leasing charge which reflects the cash flow gain from using the writing down allowances at the first opportunity. The bank or banking partnership recovers its costs and makes its profit by either taking a turn on the rate of interest included in the leasing charges, or by discounting the value of the depreciation included in the leasing charges, or by a combination of both. The level of the charges is a commercial decision which depends on the state of the market.

SUMMARY

Before searching for ship finance, it is most important that the prospective purchaser or lessor and the builder are well prepared and know exactly what they are building and what they need to have to build the ship. The financial and technical analyses must be well thought out and be thorough in their application, if they are to be accepted without major revision by the SMFC.

In the case of concrete ships there will be additional psychological problems to overcome because of the lack of precedent. Overcoming this will be time-consuming and, at times, frustrating, but no doubt success will eventually occur.

THE FUTURE

Technical developments in shipping are occuring very quickly. The container revolution is now well known; we are now facing the changes brought about by the barge-aboard-catamaran (BACAT) and lighter-aboard-ship (LASH) systems, where specialized barges are used to transport goods from an inland site to a similar one in another country. Inland, the barges are floated by canal and on arrival at the docks are loaded onto, or are unloaded from, a specialized ship. It is a quick and simple method of carrying large quantities of goods with a minimum of trouble and congestion.

Concrete afloat. Thomas Telford Ltd, London, 1977, 105–112

Prestressed concrete and related technologies have freed the naval architect from the constraint of past shipyard practice, to concentrate on configurations suited to the intended uses and the environmental conditions under which the vessel will serve. This makes possible more efficient concepts and configurations. Design criteria are being developed to serve as a basis for rational design. In addition to the traditional concerns of hog–sag moment and hydrostatic pressure, guide-lines are being developed for shear, impact, thermal strains and fatigue. Design techniques include use of unstressed and unbonded tendons in addition to the primary stressed tendons, and use of well distributed mild steel reinforcement, welded wire mesh and confining stirrups. Construction procedures especially suited to concrete vessels include methods of fabrication; cast-in-place concrete using panel forms or slip forms, and precast segments; special methods of launching (basin, sand dock and incremental launching) and completion afloat (as by rotation and jointing of modules). New materials under development may revolutionize the development of concrete ships. Very high strength concretes, lightweight concrete, fibres and polymers offer enhanced performance and economy. Prestressed concrete vessels are especially well suited to special cargoes, including cryogenic gases and hazardous materials, and offer increased assurance of safety under service and accident conditions. The optimum development of concrete ships and floating structures will take place through the full integration of design and construction.

Current trends in design and construction of concrete ships

B. C. GERWICK, Jr*

The current world-wide surge of interest in the utilization of prestressed concrete for the hulls of ships and floating structures is primarily the synergistic result of two forces: first, the needs of new industrial facilities in the ocean, involving unconventional configurations and special properties; and second, the dramatic development of new technology for design and construction in prestressed concrete.

There are, of course, other motivations, such as availability and transient economic situations. However, these serve only to initiate new projects and cannot be the basis for a long term viable industry. The demise of the concrete ship programmes after World Wars I and II proved this lesson.

Applications of current interest and importance are floating offshore production facilities for oil, gas and minerals; floating terminals (Fig. 1); floating process plants (Fig. 2); offshore storage; floating power plants and ships for transport of special cargoes such as cryogenic gases or aggressive chemicals. Environmental influences include the shallow, hot and salt-saturated seas of Arabia, the hot humid tropics of Brazil, the extreme storm conditions of the northern North Sea, and the ice sheets and pressure ridges of the Arctic Ocean[1] (Fig. 3).

---

* Professor of Civil Engineering, University of California, Berkeley.

Fig. 1. Arco LPG floating terminal for Java Sea, designed and constructed by Transenergy Inc. and Concrete Technology Corporation

These applications and environments require a re-examination of our basic approaches to design and construction.

DESIGN

The process of design is first of all conceptual, to meet needs and resist forces; and secondly, analytical, to ensure proper behaviour of materials under these imposed forces.

The conceptual aspect is, I believe, of greatest importance. Prestressed concrete has freed us from the restraints of conventional shipyard practice and allowed us to concentrate on in-service performance. With concrete, we can build an almost unlimited range of configurations, sizes, shapes and curvatures. With prestressing, we can build in deformations and stresses to more favourably resist the forces which the vessel will encounter. With reinforcement of various types, we can implant ductile behaviour and a non-catastrophic or self-limiting failure mode.

Configurations may be selected by the naval architect which will optimize the handling, processing, and transfer of product and the performance in the ocean environment of waves and, sometimes, ice. Vessels to be permanently moored may be shaped appropriately; e.g., round, with little restriction on draught, for deep-draught floating caissons. The ability of concrete to efficiently resist high membrane compression forces can be utilized. Where the vessel must resist moving ice, both normal and inverted cones become practicable and economical, and the mass of the

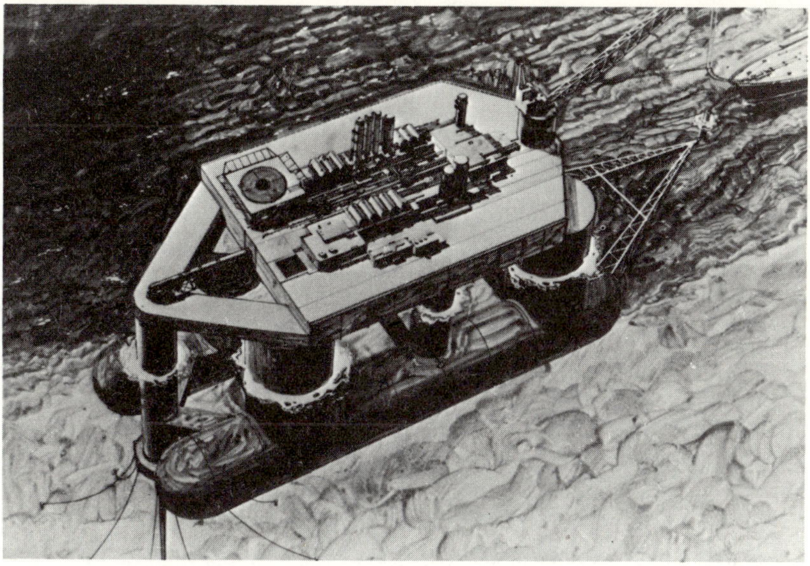

Fig. 2. Semi-submersible LNG process, storage and transfer facility (Transenergy Ltd)

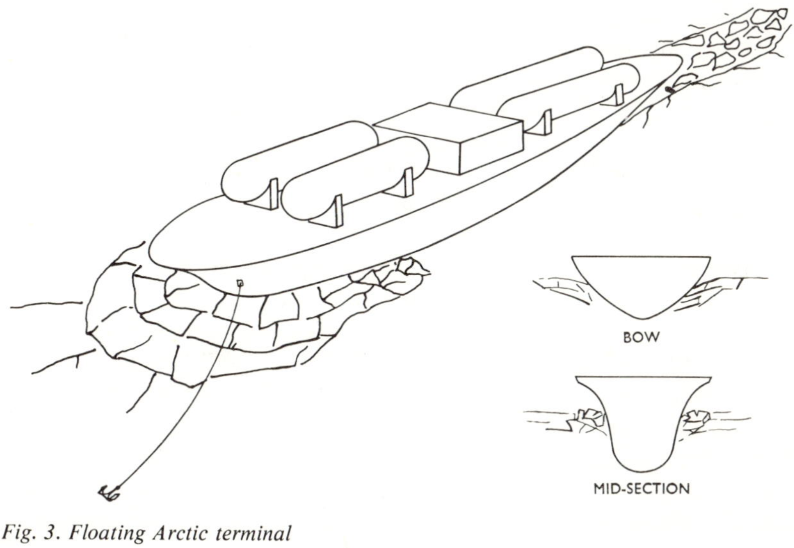

Fig. 3. Floating Arctic terminal

concrete vessel may be distributed and utilized to break the on-coming pressure ridges. For vessels which are to be used for transport, the adverse high dead weight to displacement ratio of concrete vessels may be offset wholly or in part by utilizing a more favourable shape, taking advantage of concrete's ability to be moulded into complex shapes of double curvature.

Design also includes the analysis of the adopted hull structure and the verification that the stresses and strains will be within allowable limits for the materials selected and for static and dynamic loads of random cyclic application.

Design criteria are under development by the several classification societies: Lloyds, Det Norske Veritas, American Bureau of Shipping and Bureau Veritas; also by professional societies such as the Fédération Internationale de la Précontrainte[2]; and with participation by regulatory bodies such as the United States Coast Guard.

The 1975 symposium on concrete ships, held at the University of California at Berkeley, constituted an admirable presentation of the state of art in design, construction, classification and utilization of concrete ships, and its proceedings[3] represent the most up-to-date compilation of this current body of knowledge.

Basic design criteria have been evolved to cover the principal concerns of hog–sag moment under long term cyclic loading. Hydrostatically imposed forces are effectively resisted by shell configuration, or by haunched slabs with eccentric prestress. Current studies are developing more precise guide-lines on shear, including shear compression, and the effect of cracking on ultimate strength; fatigue, especially after accident or cracking; impact under normal conditions such as docking, or under abnormal conditions such as collision; and response to thermal gradients, especially those imposed by the cargo. Careful consideration is being given to possible scale effects of extremely large concrete structures, since during the various construction stages, secondary phenomena such as heat of hydration expansion and subsequent cooling, shrinkage, creep, absorption and differential elastic shortening may later affect the structural integrity and performance of the completed vessel.

These intensive efforts are not being performed in a vacuum or as an isolated endeavour, but rather are based on the extensive experience and tests with prestressed concrete in other applications such as nuclear reactor pressure and containment structures, long span bridges, offshore drilling, production and storage platforms, and industrial applications. Even the tests and experience in cyclic loading with the lowly prestressed concrete railroad sleeper have proven of relevance. Therefore, the design criteria and practices for concrete ships should be viewed as a continuum with the other recent striking developments in the use of prestressed concrete.

Techniques and details which can be effectively utilized in the design of concrete vessels include the use of mild steel continuity to ensure ductile post-cracking modes of behaviour. Welded wire mesh may be used to control the cracking pattern and crack width under excessive strains due to accidental release of a cryogenic cargo or collision.

Heavy confinement of compression members and joints will permit increased ultimate tensile capacity, impact resistance and curvature (ductility).

Fatigue may be controlled by limiting stress ranges in normal service and by providing the welded wire mesh referred to above plus unstressed or even unbonded prestressing tendons designed to function in the post-accident condition. These are especially effective in liquefied natural gas (LNG) applications because of their relatively unchanged properties at cryogenic temperatures.

Impact may be similarly controlled by the use of mesh and unstressed/unbonded tendons. Provided post-cracking ductility is ensured, such heavily reinforced concrete has excellent properties of energy absorption and crack propagation resistance.

CONSTRUCTION

Construction methods are the key to quality, economy and schedule. The unconventional configurations and large sizes demand new construction concepts. The need for high strengths, uniform quality, and close tolerances dictate special fabrication procedures and control.

Construction aspects may be divided into three sub-items: fabrication, launching and completion afloat.

Available techniques suitable to fabrication of concrete ships include cast-in-situ concrete with panel forms, and slip-forming of constant vertical cross-sections, or, as a result of new developments, of tapered or even inclined cross-sections. Precast segments may be cast in their most favourable attitude, in a factory or on site, and then assembled with cast-in-situ joints, or matching 'epoxy' joints.

Precast segments can quite readily be cast to complex shapes, involving double curvature. Use of sand moulds or flexible adjustable forms permits the attainment of any desired shape. One of the more challenging construction details is the provision of mild steel continuity across joints: a number of schemes have been devised for this.

Launching of these large massive vessels present a problem generally exceeding the capacity of typical shipyard slipways. One very practicable method is to build the lower hull in a graving dock or basin, then launch it by flooding the dock. This is the process used for the offshore concrete platforms of the North Sea.

Another method, potentially more economical, is the use of the sand dock. For this, the basin is dredged but left full of water. Then it is filled with sand. The vessel is now fabricated at yard level. When structurally complete, it is launched by controlled excavation of the sand, using sand pumps and jetting. The settlement of the vessel is monitored by direct survey and strain gauges. After the vessel is launched, the sand is pumped back into the basin for the second use.

A third attractive method, needing development, is incremental launching, similar to the Takschiebeverfahren process for long span bridges, in which the structure is progressively cast in one casting position, then jacked forward over a pier, using low friction bearings. The difficulty in applying this process lies in adjusting for variations in buoyancy, including those due to tide: one possibility is the use of temporary buoyancy tanks on the outer end with the provision of temporarily free-flooding compartments, and controlled ballasting as the vessel's length is extended.

Concrete vessels may be constructed in large increments (e.g., halves or quarters), each being launched separately, then all joined afloat. This is of course standard practice with large steel supertankers. It has been applied successfully to the floating pontoons of the Hood Canal Bridge in the State of Washington, USA, to establish structural units of 400 m length.

Jointing mechanisms include post-tensioning, plus the use of cast-in-situ concrete or epoxies.

Concrete vessels can be built in an attitude other than their final aspect, and then rotated, either vertically or horizontally, to their correct aspect. This is more readily done with concrete than steel because of the usually adequate hydrostatic pressure-resistant qualities inherent in most concrete structures, enabling a wide range in draughts during construction. Once the basic hull structure has been rotated and jointed, completion afloat resumes the basic fabrication approaches. The partially completed hull may be used as a work platform to facilitate its own completion.

It is important to emphasize that a thorough analysis must be made of the stresses and deformations for each stage of construction; in many cases, critical stages will occur then. Any cracks occurring during construction may, depending on their location and width etc., reduce the service performance or durability of the completed vessel.

MATERIALS

The major efforts currently being made at this and other symposia, in professional and classification societies, and by individual firms is justified not only because of current potential applications, economy and safety, but also with an optimistic eye to the future. There are revolutionary developments in the laboratories today, working with new concretes of significantly enhanced properties which have special importance for concrete ships.

High strength lightweight concrete was utilized in the Genoa floating dock, and is planned for use in the Global Marine floating LNG plant. Its use is particularly valuable when the special properties of light unit weight (less than two thirds that of conventional concrete when submerged), low thermal conductivity and reduced modulus can all be utilized, as indeed appears possible with cryogenic cargoes.

Very high strength concretes are now practicable, using the new super water-reducing admixtures. These lower the water requirement while maintaining workability, thus enhancing compressive and tensile strength, as well as impermeability and durability.

Steel fibres substantially improve the apparent tensile strength of concrete, a fact which is very significant under current design criteria, where tensile strength under the extreme hog–sag wave controls the design and hence the concrete hull plate thicknesses and weight. Steel fibres are being currently used in other applications with many of the earlier problems of 'bunching' and orientation now solved. However, final failure of typical fibre concrete in tension is due to bond slippage and has a brittle character.

New 'headed' fibres being marketed in Australia appear to prevent this bond failure, giving a ductile, tough quality as the fibres elongate in tension. They also are reported to be capable of machine mixing, without bunching.

Polymer concretes are under intensive development in the laboratory and field to solve problems such as the corrosive effect of de-icing salts on bridge decks. These show greatly improved compressive and tensile strengths as well as enhanced impermeability and ductility. They too suffer a brittle failure mode. However, recent research in the UK has shown the very attractive combination of polymers plus fibres to have great ductility and toughness.

SAFETY

Of increasing importance with hazardous cargoes is the public concern with safety, under all credible service and accident conditions. Properly designed and constructed vessels of prestressed concrete have a major contribution to make in ensuring increased safety. This is due to the composite nature of the concrete itself plus the multi-axial reinforcement, the undiminished ductility under low temperatures, the fire endurance, and the redundant stress paths possible with shell configurations, two-way slabs and continuity.

CONCLUSION

The oceans are once again assuming a dominant role in mankind's advances, this time not only for communication and transport, but as a source of energy and even food. These new roles require special-purpose vessels, vessels having configurations adopted to the service use and their environment.

New configurations and shapes are made possible by prestressed concrete. Design procedures are ensuring better performance in service and greater safety in accidents. New construction methods enable the completion of these unconventional structures with the assurance of high quality and economy. New materials are emerging which will constitute a breakthrough in the previous barrier of high weight and which will enhance performance and economy.

The onward development of concrete ships and floating structures depends on the continued integration of design and construction, since these two aspects are intertwined to a high degree in these exciting new applications of prestressed concrete and naval architecture.

REFERENCES

1. GERWICK B. C., Jr. Prestressed concrete floating terminal for Arctic Ocean service. Proceedings, third conference on ports and ocean construction under Arctic conditions, Fairbanks, Alaska, 1975. (Also, *Ocean Industry*, 1975, Nov.)
2. FÉDRATION INTERNATIONALE DE LA PRÉCONTRAINTE. *Concrete ships: guidelines for design*. FIP, Wexham Springs, Slough, 1975.
3. GERWICK B. C. (ed.). *Proceedings, conference on concrete ships and floating structures*. University of California, Berkeley (Extension Division), 1975.

Concrete afloat. Thomas Telford Ltd, London, 1977, 113–119

An outline is given of the various functions of marine surveyors, including classification societies, salvage associations, governmental authorities and a marine surveyor acting as a result of an insurance warranty. Typical recommendations which may be made by a marine surveyor for a floating concrete unit are discussed. Other major factors which a marine surveyor may be requested to consider are also outlined.

The role of the marine surveyor

B. P. M. SHARPLES, BESc, PhD, PEng, ACSCE, AMEIC*
S. M. TAIT, BSc, MICE, MIMunE†

A marine surveyor's primary function is to observe independently and report impartially on his client's marine activities.

Traditionally, the name 'marine surveyor' has been associated with cargo surveying, on and off hire surveys and ship damage surveying. Owners, operators, insurers and financiers all may require an independent opinion on the condition of a vessel or other floating unit, or a cargo, at different points in time. Marine surveyors are also often requested to give independent opinions on the suitability of floating units for marine ventures.

Classification societies are perhaps the best known of all surveying authorities and such names as Lloyds Register of Shipping, Det Norske Veritas, American Bureau of Shipping, Bureau Veritas, Germanischer Lloyd and others are world renowned in shipping circles. Perhaps an owner's best safeguard in assuring himself of the quality, design specification and function of any marine equipment he may be constructing is to have the equipment built to class or under the supervision of a classification society. Their function is generally to give owners, contractors, national governments and underwriters a valuable second opinion on the design and construction of a marine unit. Such objects may in many cases be of advanced design created by engineers, naval architects and consultants pioneering in their disciplines and working with new materials, often related to an unfriendly and largely unknown environment.

The Salvage Association in London and the United States Salvage Association, as their names imply, are mainly involved in the assessment of marine casualties on behalf of interested underwriters. This work consists of surveying and quantifying the extent of damage sustained, of assessing the repair and salvage costs (if repair and salvage are possible),

* Vice President, Noble, Denton and Associates Inc., Houston.
† Project Manager, Noble, Denton and Associates Ltd, London.

of advising underwriters of their opinion and of exercising control of repairs and salvage costs in underwriters' interests. They also operate on behalf of clients in the traditional role of the marine surveyor.

Government marine safety authorities are concerned with maintaining and establishing the safety of personnel working in the marine environment. They are responsible for administering their own government and international marine safety laws regarding the stability of manned floating equipment, life-saving provisions, fire-fighting appliances, navigation lights, shapes and signals etc. They are also responsible for administering the international loadline regulations and tonnage laws.

The authors' company carries on the normal functions of marine surveying houses but has specialized, among other things, in advising on towages of an unusual nature and entailing a high marine risk, and in the operational aspects of mobile offshore barges and platforms. The company is often required to furnish owners with recommendations and approvals for the above activities as a condition of the insurance.

The organisation, under the name(s) of its principal officers, has been providing approval services to the London insurance market since before World War I. Initially, involvement was primarily with dredgers, damaged or scrap ships, cranes and floating docks. In the 1950s this activity was extended to cover the large range of units required for oil field exploration and development.

Approvals are provided for ventures insured on towage, operations and builders' risk policies placed in the London and overseas markets.

Notes on the marine surveyor's review procedure are given in Appendix 1 for an ocean towage and in Appendix 2 for the construction of a concrete structure.

Experience has shown the usefulness of preparing a curve of critical motion for towage of a unit. The curve is individually prepared for each unit when in its floating condition and indicates allowable angles of heel against periods of roll and/or pitch for the critical stress condition of the unit, its cargo or sea-fastening. Reference to this curve would indicate to the personnel riding or in charge of the tow when the motions are reaching an undesirable level and when positive action, such as changing course to alter angle of wave attack or changing speed, should be taken.

In units where the forecast motions are small, it may be necessary to provide instrumentation to monitor motions which can be compared with critical motion curves. For barge tows, simple inclinometers can be used with or without recording facilities, depending on whether the prototype information is to be used in future designs. In deep-draught floating concrete units, accelerometers may replace inclinometers and the critical motion curves may be replaced by limiting accelerations which will combine static and dynamic forces if surge forms a large proportion of the total motion. If these accelerometers are installed in similar positions to those on the model, results can be compared.

Model tests form an increasing part of the design and development of ideas for marine transportation. Consequently the marine surveyor spends some of his time in assisting with the undertaking of such tests in such aspects as advising on the environmental criteria to be considered, assessing the results of the tests, advising on marine operations which could reduce the motions of the unit and advising on the design of the unit regarding its stability, nature of sea-fastenings etc.

This concludes a brief summary of the many points to be considered in approving construction, towage and emplacement of concrete and other structures.

APPENDIX 1. TYPICAL JOB INVOLVEMENT: OCEAN TOWAGE

Outlined below are subject headings and points which may be reviewed by a marine surveyor when assessing the suitability, for example, of a concrete floating unit departing on an ocean towage. Brief comments and recommendations are included where these are of a general nature and appropriate.

Documentation

- Certificate of ownership
- Customs clearance
- Deratization certificate
- Certification of navigation lights and shapes
- Certificate of class
- Loadline or loadline exemption certificate
- Certificate for fire-fighting and life-saving appliances
- Tonnage certificate (incorporating canal tonnage, if required)
- Canal transit authorization

Towing vessels

- Documentation
- General arrangement and suitability
- Motion characteristics
- Capacity of tanks
- Details of any special characteristics such as ice classification or any other geographical limitations
- Machinery and fuel consumption
- Fuel capacity and effective horsepower
- Refuelling programme
- Details of arrangements for standing by and controlling tow in the event that the towing vessel is required to leave for any purpose

Static and dynamic stability characteristics

Full details of towing arrangements, including towing winches, number and sizes and condition of main towing wires and chains, towing pennants, spring lines etc., spare main towing wires with details of stowage and spare towing chains, pennants, spring ropes, shackles, thimbles, and soft cordage required for handling towing equipment

Details of arrangements and equipment to be fitted or available on towed unit to allow recovery of towing equipment in the event of a towline breaking or for other reasons

Manning, including qualification and experience of intended crew and any arrangements for placing a riding crew on tow

Communications systems, including arrangements for weather forecasts

Workboats available for transferring men and equipment from tug to towed unit in an emergency during the towage

Details of any equipment available on board the towing vessel, such as portable pumps with suction and delivery hoses, air compressors, portable generators for transfer to towed vessel if required

Navigation charts and electronic navigational systems, echo sounders etc.

Manning

Experienced riding crew capable of maintaining machinery, carrying out temporary repairs, inspecting sea fastenings, communicating accurately with towing vessels the motion and condition of the vessel in a common language, retrieving and reconnecting the towing arrangements

Adequate accommodation, messing and sanitary facilities, domestic fuel and potable water

Breakwater

A breakwater may need to be erected to protect accommodation and plant equipment and deck cargo not designed for wave slap, and personnel who may be reconnecting towlines, etc. Guard rails to be provided around periphery of barge.

Emergency anchor(s)

Emergency pumping arrangements

All tanks should be capable of being pumped out either by installed system or by portable pumps. Manholes or equivalent access should be made to all compartments. Portable manhole coaming (top hat) available, depending on freeboard, to be provided in order that any tank may be pumped dry with water on deck.

Sounding arrangements

Sounding arrangements to be provided for all compartments.

Watertight integrity

All penetrations through watertight bulkheads (including doors) to be satisfactorily sealed. Accommodation to be watertight. Manhole covers to be properly secured watertight, bolts and gaskets renewed as necessary. Control of all suction and discharge apertures in hull.

Securing of equipment

All equipment to be stowed satisfactorily against lateral and vertical movement on and below decks. High-standing pieces of equipment to be adequately secured to withstand forces due to roll, pitch, heave and surge, or stayed to dampen or break up synchronous motion. All pipework, cables, trays to be well clipped and protected from 'green seas'. Tie-down to be checked for static, dynamic and fatigue forces.

Stability

Adequate static range of stability
Dynamic range with adequate factor of safety
Effect of ballast and free surface
Consideration of draught and trim

Damage control equipment

Strength

Motion response

Rolling, pitching, heave and surge to be studied in conjunction with model tests. Generally a concrete barge, being heavier than a steel barge of the same capacity, may have a lower centre of gravity if the barges' waterplane areas are similar. This in turn would imply a higher metacentric height which may result in greater accelerations at deck level.

Route/weather services

The proposed route and possible ports of shelter or standby areas to be discussed with master of towing vessel. Certain tows may require the specialist advice of weather ship-routing organizations.

Helicopter deck

Cranes

Tie-down to take account of unit motion and flexure.

Ancillary equipment

Adequate means of illuminating the deck and its equipment to be provided.

Chain of command

Firmly establish who is in command, whether the master of the tug or the master of the vessel being towed.

Electrical equipment

All electrical equipment to be commensurate with a seagoing vessel to be satisfactorily protected aginst weather damage.

APPENDIX 2. TYPICAL JOB INVOLVEMENT: BUILDING, MOVING, MOORING AND CONSTRUCTION AFLOAT

The construction of a concrete structure may involve building the unit in a dry building basin, floating it out in pieces, joining it afloat, and/or mooring it during construction afloat. During these phases sufficient detail should be paid to the following.

Cofferdam being constructed to prevent overtopping in storm conditions
Precautions against grounding during exit from building basin to open sea
Assessment of tides from tidal predictions compared with observed tides for the location, and observations of barometric pressure and other local conditions affecting water level related to a common datum
Securing of unit inside basin during removal of cofferdam
Prevention of premature lift-off by ballasting during removal of cofferdam if unit does not float at any state of the tide
Adequate deballasting arrangements to ensure sufficient under-keel clearance
Control of unit when leaving basin and room for manoeuvrability
Suitability of tugs and winches for moving out of building basin
Stability, damage stability and strength of unit in uncompleted states and in relation to any ballasting during construction
Adequate lighting if move needs to take place at night due to local tide conditions
Channel markings
Temporary pumping arrangements in event of accidental leakage
Definition of weather criteria for float-out and joining together if segmented construction used
Control and communications during movements
Adequacy of temporary moorings to resist extreme storm conditions including sea bed conditions, pretension devices to reduce excursions and connection and disconnection procedures
Commissioning of ballasting systems

Temporary moorings and fendering for construction craft alongside unit

Marine operations related to skidding out or lifting packages onto unit

Adequacy of freeboard during construction afloat to prevent wave overtopping

Adequate number of suitably strong towing connections for the maximum-sized towing vessel

Watchkeeping

Contingency plan availability of damage control equipment on board and back-up supplies

BIBLIOGRAPHY

WERENSKIOLD K. Maritime operations relative to construction of large concrete offshore structures. *Design and construction of offshore structures.* Institution of Civil Engineers, London, 1977, Paper 10.

NOBLE-SMITH D. R. The role of the surveyor in North Sea hydrocarbon fields. *Petroleum Times*, 1974, Oct.

NOBLE W. D. Towage, insurance and safety. *Symposium on offshore drilling rigs.* Royal Institution of Naval Architects, London, 1973.

DENTON A. A. Anchoring, towing and installation of structures. *Proc. Instn Civ. Engrs*, Part 1, 1976, **60,** May, 255–257.

FÉDÉRATION INTERNATIONALE DE LA PRÉCONTRAINTE. *FIP recommendations for the design of concrete sea structures.* FIP, Wexham Springs, Slough, 1973, first edn.

Concrete afloat. Thomas Telford Ltd, London, 1977, 121–130

Though concrete ships will be designed from first principles, using to full advantage the properties of concrete as a shipbuilding material, a direct comparison with the performance of steel ships is unavoidable. Structural design criteria for steel ships are briefly presented, indicating safety margins available at extreme load conditions. An approach for structural appraisal of concrete ships is outlined. A differentiation is made between permissible-stress approach as practised today in ship structural design and a limit-state approach recently adopted by civil engineering industry.

Structural appraisal of concrete ships

B. RAPO, diplIng, MRINA*

The use of prestressed, reinforced concrete in commercial shipbuilding has hitherto been of rather academic interest, particularly insofar as large, ocean-going ships are concerned. However, recent developments have brought about a change in this situation.

It is often said that, in general, the design of ships, as practised by naval architects, is an exercise in both art and science. The design of prestressed concrete ships adds a new dimension to this. It is quite clear that the simple marriage of existing ship design principles and current prestressing building techniques, however well tried in a non-marine environment, will not suffice in the building of technically sound and commercially viable ocean-going vessels. The successful design must embrace all the constraints imposed by marine and civil engineering, concrete material technology, operating requirements and existing statutory regulations, fatigue environment and also ecological and environmental protection considerations. In some of these aspects, the design of large ocean-going concrete vessels requires the extrapolation of existing knowledge and experience.

It would appear that, due to the less favourable dead weight/displacement (dwt/Δ) ratio in concrete ships, the main area of their application will be restricted, at least initially, to capacity carriers; i.e., the vessels which are intended for the carriage of low density cargoes; e.g., liquefied natural gas (LNG) ($\gamma = 0{\cdot}45$), possibly ethylene ($\gamma = 0{\cdot}57$) and propane ($\gamma = 0{\cdot}583$). There is a wide variety of vessels for which a low dwt/Δ ratio will not prove to be a restriction on operational profitability; e.g., floating docks, mobile dry dock facilities for maintenance of steel semi-submersible rigs, all types of stationary vessels and cargo handling installations, permanent storage facilities, towing barges, dry cargo ships intended for special light density cargoes, and stationary re-liquefaction platforms.

* Senior Surveyor, Lloyd's Register of Shipping, London.

The reasons for the selection of concrete as a building material in these applications may be different in each case. Specific reasons could be low maintenance costs, or the long life expectancy of concrete structures, or comparative savings in initial costs etc. The cryogenic potential of concrete should not be disregarded. A number of shore LNG storage facilities, with the concrete permanently exposed to a cold environment, are now in full operation, particularly in the USA and in W. Germany. In some instances, LNG at −160°C is stored with only a liquid-tight plastic membrane liner interface separating the LNG from the concrete surface. There are many marine applications where the cryogenic properties of the concrete can be utilized to the full, especially when the temperature gradient between the inner and outer face can be controlled successfully.

To add a historical note to this, Lloyd's Register's involvement in classing reinforced concrete ships can be traced as far back as World War I, when an anticipated steel shortage caused the shipping world to turn to concrete as an alternative to steel. Amongst the first entries for reinforced concrete ships in the register book was m/s *Polias* (81·4 m × 14·2 m × 11·4 m, $d = 6{\cdot}9$ m). The ship was completed in early 1919.

More recently Lloyd's Register of Shipping completed an initial feasibility study including plan approval and appraisal for a large prestressed and reinforced concrete LNG carrier of 125 000 m^3 capacity.

Whilst it is true to say that a correct approach to the problem of structural design of concrete vessels would be based on the first principles of engineering specially adapted to a building material which is novel as far as the steel shipbuilding industry is concerned, it is impossible to discuss the subject at this stage without making continuous reference to steel ships.

STEEL SHIPS—STRUCTURAL DESIGN CRITERIA

Steel ships are designed to safety standards formulated by the classification societies. In most cases it will be found that a proper combination of all static and low frequency primary, secondary and, where necessary, tertiary stresses will result, after allowing for stress concentrations in critical regions, in a maximum total stress approaching the yield point of the steel material used. The required margin below the yield point is in most cases dictated by fatigue considerations, but the buckling criterion can become a governing factor.

Lloyd's Register of Shipping requires that the design midship section modulus, $(I/y)_m$, should not be less than

$$C_1 L^2 B (C_b + 0{\cdot}7) \text{ cm}^3$$

where

I is moment of inertia of midship section;

y is distance from neutral axis;

m denotes minimum requirement;

C_1 is factor determined as a function of ship's length, given in Table D3.2 of the Society's rules;

L is length of ship, m;

B is breadth of ship, m;

C_b is moulded block coefficient.

The maximum permissible still-water bending moment M_s associated with $(I/y)_m$ is given by

$M_s = 8{\cdot}15 \times 10^{-3}\ (I/y)_m$ t m for types 2 sea-going vessels

and

$M_s = 6{\cdot}4 \times 10^{-3}\ (I/y)_m$ t m for type 1 sea-going vessels

The definition of ship types is given in para D.318 of the rules. The coefficients 8·15 and 6·4 in the expressions for M_s are the maximum allowable bending stresses at the deck and bottom shell for types 1 and 2 respectively, in kp/mm², all in still water.

A rigorous determination of the vertical wave bending moment M_w and of any other wave load components is very complex. During the past decade or so there has been a rapid advance in the theoretical study of ship behaviour in open seaway. The deterministic approach to longitudinal bending has been replaced by a statistical approach based on spectral analysis. The evaluation of individual wave loads and motions can be carried out using the Society's suite of computer programs within the LR SHIP system, based on regular response transfer functions obtained from strip theory.

For the convenience in structural design, the Society has incorporated in the rules an empirical expression intended to be used as the design wave bending moment, M_w:

$$M_w = 10\, C_1\, L^2\, B\, (C_b + 0{\cdot}7)\, 10^{-3}\ \text{tm} \qquad (1)$$

The maximum bending stress at the deck and bottom shell caused by M_w, as given by expression (1), should not exceed 10 kp/mm².

Whilst the sum of bending moments $(M_s + M_w)$ can be taken to represent the primary static and environmental 'demand', the structural 'capability' is determined by the load-carrying ability of the structure associated with permissible stresses, as the main, primary response parameters:

$\sigma_c = \sigma_s + \sigma_w = 16{\cdot}4$ kp/mm² for type 1 ships

$\sigma_c = \sigma_s + \sigma_w = 18{\cdot}15$ kp/mm² for type 2 ships

CONCRETE SHIPS—STRUCTURAL DESIGN CRITERIA

Rational structural design criteria for concrete ships must necessarily reflect both the current procedures employed in the design of steel ships, largely governed by requirements of the classification societies, and the limit-state design approach recently adopted by the civil engineering industry and expounded by various structural codes.

The permissible-stress approach (PSA), the basic criterion in steel ship structural design, whilst ensuring satisfactory performance of the vessel under maximum design loads, has the limitation of not providing a direct and true indication of the margin of safety against failure.

Limit-state approach (LSA), which replaces a single failure criterion, controls the serviceability of the structure and its safety at all stages of its service performance. It incorporates the following criteria:

(*a*) limit states of serviceability, to ensure satisfactory performance under working loads;

(*b*) ultimate limit state to ensure that the probability of failure is adequately low.

Limit-state design is based on partial safety factors which are applied separately to static and environmental loads (γ_f) and material strength (γ_m). The logic behind this is that the γ_f factors cover uncertainties in the assessment of the loads and the γ_m factors are intended to cater for uncertainties in the performance of the material. The overall safety factor is then equal to the product ($\gamma_f \gamma_m$).

Permissible-stress approach

The permissible-stress approach requires complete determination of all load components which affect longitudinal, transverse, torsional and the local strength of the vessel. These may be given as follows.

Static loads:	M_s^c	— still-water bending moment
	M_t^c	— post-tensioning bending moment
	SF^c	— shear force in still water
	F_t	— post-tensioning axial force
Dynamic loads:	M_w^c	— design vertical wave bending moment
	H_w^c	— design horizontal wave bending moment, where relevant
	T_w^c	— hydrodynamic torque loading, where relevant
	SF_w^c	— vertical wave shear force
Local loads:	d	— design load draught
	d_b	— design ballast draught
	d_{ud}	— assumed draught to upper deck
	h_w	— hydrodynamic wave pressure head
Test loads:		— test head
Accidental loads:		— collision
		— grounding
Thermal loads:		— where relevant

Regarding the relative magnitude of the above load components when compared to equivalent steel ships, the following comments can be made.

M_s^c—all loading and ballast cases should be examined and an envelope curve, embracing all still-water bending moments, established; M_s^c is to be taken as the maximum ordinate of the envelope curve;

M_w^c—in most cases, the design vertical wave bending moment for concrete ships can be obtained using the expression given for steel ships; if necessary it can be vertified using the long term estimate of responses;

F_t—as a difference from steel ship construction, the prestressed concrete hull will be brought to a permanent state of compression prior to launching or floating out, whilst in the building dock and upon hardening of the concrete; the required amount of post-tensioning should be specially determined in each individual case as a function of the total design bending moment ($M_s^c \pm M_w^c$) in hogging and sagging condition, the geometrical characteristics of the cross section, and the permissible stress/strain levels;

M_t^c—only in exceptional cases will the centre of application of F_t coincide with the centre of gravity of the cross section. In the majority of cases the designer will use to advantage the post-tensioning bending moment created by the eccentric application of F_t.

Flat or arched panels spanning in two directions will be exposed to simultaneous longitudinal and transverse membrane forces and local lateral loading due to the buoyancy, cargo pressure, own weight, the weight of water ballast etc. A method of designing the reinforced and post-tensioned sections subjected to biaxial bending, membrane and shear forces should take account of all stress components as shown in Fig. 1.

The determination of axial, bending and shear stresses, for a design method based on the 'complete freedom from cracking at maximum design loading' concept, is based on the application of the beam theory and elastic flexural analysis. This analysis requires determination of the net, gross and effective gross properties of the midship section.

The net properties of the midship section are calculated for the non-grouted cross section, which exists after hardening of the concrete but prior to the application of post-tensioning. These properties are used to evaluate the axial and bending stresses in the concrete and steel reinforcing bars caused by the application of post-tensioning.

The gross properties of the midship section are calculated using the entire net concrete section including grouted ducts, plus the steel reinforcements suitably factored on an E_{st}/E_c basis plus the area of post-tensioning tendons increased in the ratio E_{st}/E_c. These properties are used to evaluate the bending stresses due to static and wave loads which do not exceed the value of threshold bending moment.

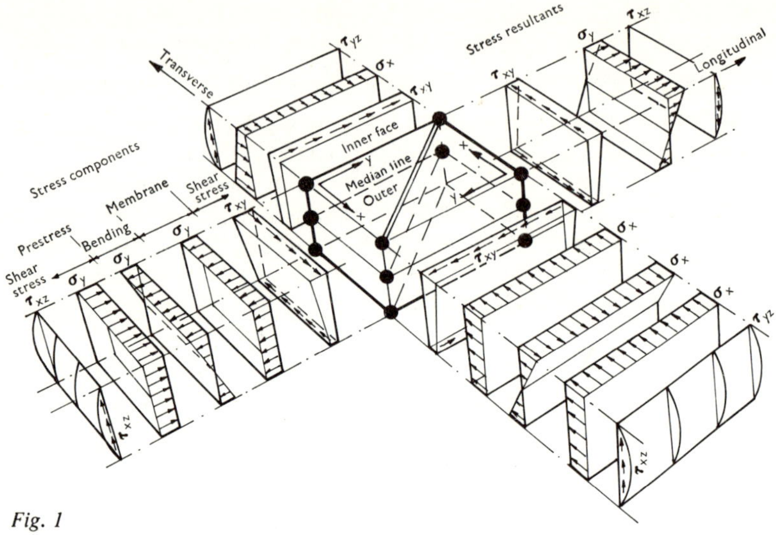

Fig. 1

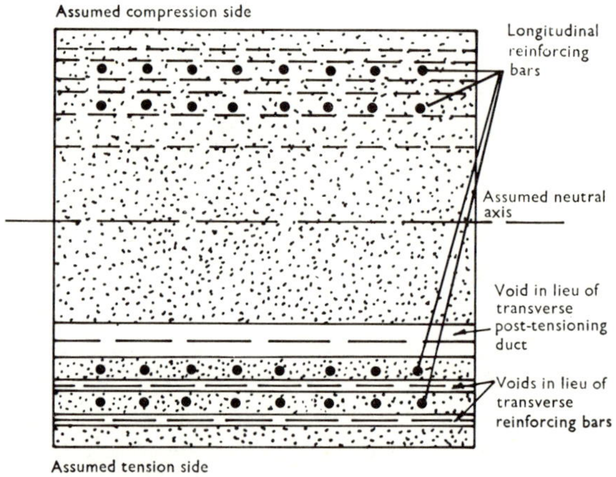

Fig. 2. Unit block: idealization for calculation of effective gross section properties

For heavily cross-reinforced concrete sections the net concrete sectional area is difficult to define. Fig. 2 shows a typical unit block with the portion below the neutral axis exposed to tension (in the out-of-plane direction), the upper portion being in compression. Transversely orientated reinforcing bars or post-tensioning ducts do not influence the behaviour of the section in compression. However, in the tension zone, the tensile resistance of the layer of material adjacent to these bars is considerably reduced and is equal to the bond stress between the steel bars and the concrete. The current practice of the Society is to ignore the contribution of these bond stresses, and hence the effective gross section properties are calculated in the same way as for the gross section properties, with the concrete area adjacent to the transversely orientated reinforcing bars ignored in the tensile zone. This leads to the situation whereby two sectional properties, hogging effective gross section properties and sagging effective gross section properties, exist for the same unit, dependent upon which portion of the section is taken as the tensile zone.

Effective gross section properties are used to evaluate bending stresses for any increment of wave loads above the threshold bending moment. The threshold bending moment is defined as that bending moment which is of sufficient magnitude to nullify the compression imposed by post-tensioning forces and, as such, will cause zero strain in the deck for a hogging ship or zero strain in the bottom for a sagging ship.

The intensity of axial and primary bending stresses normal to the section, at any given point of the cross section, is given by the general equation

$$\sigma = \frac{F_t}{A} \pm \frac{(M_s^c \pm M_w^c)}{I_x} y \pm \frac{H_w^c}{I_y} x\beta$$

where β is the phase angle correction factor.

In general, the allowable nominal stress levels depend on the ability to predict the loading and the reliability and accuracy of the methods applied in the determination of the response capability of the structure. The Society's present position is that LNG carriers should be designed on the following basis:

(*a*) there should be complete freedom from cracks in concrete under maximum design 'once in a lifetime' loading conditions (class 2 design in terms of the structural code);

(*b*) limited width cracking due to local bending (class 3 design) is allowed only in structural members of secondary importance;

(*c*) the maximum compressive stress in concrete is not to exceed 50% of its characteristic cube strength;

(*d*) the maximum tensile stress in the concrete is not to exceed 80% of its modulus of rupture or $0{\cdot}7\ f_{cu}^{2/3}$ whichever is less, where f_{cu} *is* characteristic cube strength in kp/cm^2;

(*e*) performance of the structure under combined shear loading should be shown to be satisfactory;

(*f*) there should be strict quality control, testing and inspection, associated with well defined rejection criteria;

(*g*) bonding performance of the insulation, together with its thermal and mechanical properties, is to be proven satisfactory for 20 years of simulated ship operation.

Limit-state approach

The limit-state approach requires the establishment of characteristic load components comprising service and environmental loads, partial safety factors for loads (γ_f), partial safety factors for materials (γ_m) and permissible stress limits.

The following design loads are to be considered:

(*a*) standard service loads (SSL);
(*b*) maximum design loads (MDL);
(*c*) cracking loads (CL);
(*d*) ultimate loads (UL).

The relationship between these loads is shown diagrammatically in Figs 3 and 4.

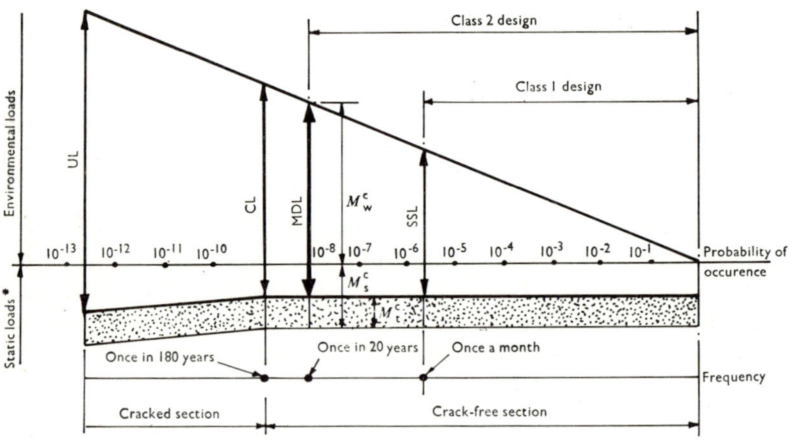

* Uniform axial post-tensioning effects should be accounted for separately

Fig. 3. Design loads for hogging condition

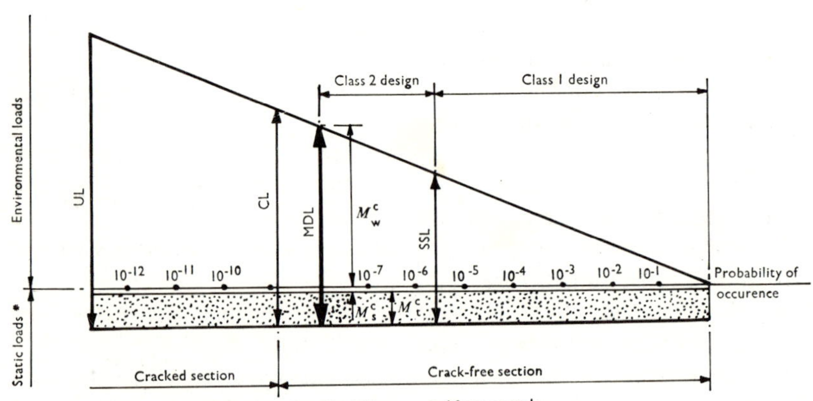

* Uniform axial post-tensioning effects should be accounted for separately

Fig. 4. Design loads for sagging condition

SSL consists of maximum static (for hogging condition) or minimum static (for sagging condition) bending moment M_s, M_t^c, and maximum environmental load occurring in one month (probability of occurrence $10^{-5.6}$).

MDL consists of bending moments M_s and M_t^c as given above plus maximum design environmental load having a probability of occurrence of 10^{-8}.

CL is a reflection of response capability. For a crack-free design, CL will always be greater than MDL.

UL is a reflection of the ultimate strength of the hull girder, which is reached when the section is about to fail. The UL is established using non-linear analysis and ultimate section design.

This approach applied to an LNG design may be illustrated as shown in Table 1, with the permissible stress in compression being obtained as shown in Table 2. It may be concluded, therefore, that the difference between the PS approach and the LS approach is superficial, the end result being similar. This is true, at least as long as a class 1 limit-state design is associated with SSL and a class 2 with MDL.

For the serviceability limit state of cracking under MDL, which may be acceptable for certain types of floating structures, the CL may well fall below MDL. In this case it is envisaged that a class 3 design should ensure that the surface width of cracks does not exceed 0·1 mm.

CONCLUDING REMARKS

Concrete construction of floating structures operating in a marine environment is in many ways a completely different proposition from that

Table 1

	Design load = design bending moment, t m		γ_f
UL	$1\,M_t^c + 1{\cdot}25\,M_s + 1{\cdot}6\,M_w$	945 000	1
CL	$1\,(M_t^c + M_s) + 1{\cdot}12\,M_w$	655 000	1·443
MDL	$1\,(M_t^c + M_s) + M_w$	595 000	1·1
SSL	$1\,(M_t^c + M_s) + 0{\cdot}7\,M_w$	445 000	1·337

Table 2

	Compound γ_f	γ_m	$\gamma_f\,\gamma_m$	Permissible stress in compression/ characteristic cube strength
UL	1·0	1·5	1·5	0·667
CL	1·443	1·25	1·804	0·554
MDL	1·588	1·25	1·985	0·504
SSL	2·124	1·25	2·655	0·3766

of dry-land concrete application. In land construction, limited cracking caused by local bending is, more often than not, an acceptable occurrence. However, ship structures operate in a more severe, fatigue environment. Membrane type cracking caused by primary bending stresses is highly undesirable, for the following reasons:

(*a*) if the crack width is greater than 0·1 mm, it is generally accepted that an inflow of sea water occurs and hence the tightness of the hull is severely impaired;
(*b*) the tensile response capability of concrete cannot be restored, since cracks will make concrete ineffective in tension;
(*c*) the stiffness of the cracked section is considerably reduced;
(*d*) the reinforcing bars would be exposed to the corrosive action of sea water;
(*e*) the long term behaviour of cracked concrete under repeated compressive loading is not sufficiently understood.

The consideration of weight-effectiveness is a factor which requires in concrete ship construction a much greater steel content, the exclusive use of lightweight aggregates and multiaxial prestressing.

The future prospects of concrete ships will be decided by the interplay between the fundamental ratios dwt/Δ and price of steel/price of concrete.

It is considered that a dwt/Δ ratio of 0·55 should be obtainable for an LNG carrier. It may be possible, for a certain size of vessel, to achieve a value of 0·58 by the combination of high strength concrete and special low density lightweight aggregates. Any further improvement of this ratio for LNG carriers, even if the draught restrictions are removed, appears to be highly doubtful.

The cost ratio steel/concrete is a powerful factor, ever changing with a marked upward tendency. Hence, the economy of construction in certain applications, which may be doubtful today, could look more attractive tomorrow.

Concrete afloat. Thomas Telford Ltd, London, 1977, 131–146

The paper outlines the technical involvement of the classification society during construction on site and subsequently, to ensure sound construction and continued integrity of the structure during its life, in order that, *inter alia*, statutory requirements for strength can be met. The interface between the field surveyors and the constructor's quality control team is discussed, together with classification approach to materials, and to maintenance and repair of these structures.

Quality control and maintenance of prestressed concrete ships

J. G. BEAUMONT*

From a structural point of view at least, a substantial number of reinforced concrete vessels up to 130 m long have proved themselves throughout significant service lives in specified operational categories ranging from canal to ocean navigation. We are now approaching what is potentially a new era. Larger prestressed concrete hull designs for specific applications are in service and others, including a 280 m long liquefied natural gas (LNG) carrier are approaching the stage where construction could commence. The economic feasibilities are being carefully considered.

The role of the classification society[1] is to formulate and maintain standards of hull structural strength, effectiveness of the main and auxiliary machinery essential to sustain propulsion, and the suitability of the arrangements for the protection of the cargo, the ship and the life on board. Such arrangements include adequate fire protection and extinction facilities and means of escape. They can be extended to cover cargo refrigeration systems. The classification certificate is virtually synonymous with the safety construction certificate (issued in accordance with the provisions of the 1960 convention for the safety of life at sea), and fully covers the strength requirement for the issuing of the international loadline certificate. Accordingly the role of the classification societies in the context of concrete ship construction is clear cut.

It is the purpose of this paper to present the approach of Lloyd's Register of Shipping to matters relating to the quality control during the construction of the hulls of prestressed concrete ships and to the procedures for their periodical survey and repair. The approach is basically one of quality assurance, whereby the surveyors keep the builders' and repairers' quality control procedures under review, and aug-

* Senior Surveyor, Lloyd's Register of Shipping, London.

ment this control by such direct inspection on their part as is considered necessary.

No attempt has been made to provide a complete coverage of the subject, but rather the intention has been to highlight those topics with which the Society and its field surveyors will be especially concerned and which might thus be of particular interest to the naval architect or to the civil engineer or both.

Classification societies traditionally operate in an evolutionary manner. Rules for new types or size ranges of ships are developed as an extension of previous experience with the specific construction materials concerned, and the rules are continuously reviewed in the light of service performance as reported by the surveyors attending on board.

Prestressed concrete is a new shipbuilding material and its use for this purpose will, by its very nature, be accompanied by a significant increase in ship size in comparison to its reinforced concrete antecedent. For this reason the Society considers it prudent to adopt a somewhat cautious approach to the classification of prestressed designs and the committee will, where necessary, assign an experimental class notation until satisfactory service experience has been gained with these ships. The early periodical survey schedule will reflect this.

At this stage we, in common with the other participating disciplines, must proceed on the best engineering principles and judgements available, neither standing in the way of innovation nor forgetting our obligation both to the cause of safety and to the good name of concrete as a construction material.

APPLICATION FOR CLASS

Traditionally the classification society is appointed in accordance with the terms of the contract between the builder and the owner. The builder is then obliged to produce a design complying with the relevant rules of the society and to obtain classification certificates on completion of construction to fulfil his contract.

Accordingly he formally requests the society to undertake the plan approval and survey during construction and is thus responsible for ensuring that all materials and subcontracted components essential to classification are ordered to be supplied with the society's certificate. Invariably the material and component manufacturers are, in turn, responsible for requesting the local surveyors to attend their works as necessary to enable such certificates to be issued.

It is to be appreciated that Lloyd's Register is capable of providing its services by exclusive representation anywhere in the world virtually independent of the size of the project. Before accepting to undertake this work, however, Lloyd's Register endeavours to satisfy itself that sufficient basic facilities and expertise are available at the proposed site to establish the feasibility of sound construction. It does not formally approve these arrangements since they are basically the responsibility of the builder. However, it does reserve the right to declare them unsuitable at any stage if they deteriorate or if a declared expansion of them does not materialize before the relevant stage of construction has been reached.

CONSTRUCTION MATERIALS

The submission of a design for approval should include a basic technical specification giving all the detail which, when taken together with the construction and other plans required by the society, will enable every aspect pertinent to classification to be properly considered. The basic technical specification may form part of the general specification agreed between the builder and owner, but it is only the former which will require the attention of the society. Accordingly the basic technical specification will incorporate complete information on the proposed materials of construction and the way in which they are to be handled at site.

Inspection procedures at the manufacturer's works will be arranged by the surveyors to ensure the consistency and acceptability of such components.

Physical properties of the concrete such as strength, workability, shrinkage, creep, elasticity, abrasion, absorption and permeability should be established at the earliest possible stage in the design. The chemical compatibility of the aggregates and cement should be similarly established. The extent of preliminary material testing will depend to some extent on the complexity and novelty of the mix in regard to marine application.

Steel reinforcement and prestressing tendons are to be tested at the manufacturer's works in the presence of the surveyors who will, on satisfactory completion of the tests, issue certificates indicating compliance with the rules.

Anchorages and couplers are to be of proven design, and tested to show compatibility of strength with the prestressing tendons. Such tests should demonstrate the physical properties of the steel itself and the performance of the device in relation to the associated standard arrangements of anti-bursting steel. Details of tendon sheaths and grout composition are to be submitted for approval.

Before construction begins, early cube tests of the concrete should be carried out at site to ensure a high level of confidence that the characteristic strength of the concrete will be achieved when construction commences.

The surveyor will review the constructor's quality control procedures for monitoring the materials entering the site, the arrangements for the subsequent storage of cement and aggregates and their mixing and handling. He will also check the histographic record of the characteristic strengths of the concrete in each primary structural zone (e.g., main deck within midship half length etc.) to ensure that the deviation from the design value does not exceed that permitted by the rules.

QUALITY OF CONSTRUCTION

Basic site facilities

The builder is responsible for providing the site facilities necessary to ensure sound construction, and these will obviously depend to some degree on the size and complexity of the ship to be built. He should first establish that the building berth foundations are suitable for the intended loadings.

The materials handling and quality control equipment will quantitatively depend on the intended pouring rates, but good storage facilities for cement and aggregates, accurate measuring and mixing equipment, adequate means to rapidly transport and accurately pour concrete at all heights and all parts of the berth, and a properly equipped laboratory are the obvious essentials. If it is not intended to mix on site but to receive ready mixed concrete from a specialist supplier, then the supplier's facilities and the distance and terrain over which the mix is to be transported will have to be taken into consideration.

Accordingly, as soon as it is announced that a particular site is to be used to build to the society's class a ship of unprecedented type or complexity, the surveyor will be looking at the facilities with a critical eye, matching them to the nature of the job in hand, to the shuttering techniques proposed and, most importantly, to both the experience of the construction and quality control teams involved and the managerial independence of those teams.

Before leaving the topic of site facilities, noting that construction will invariably take place in a building dock, it should be mentioned that the surveyor will, in general, require an examination of the soffit of the bottom shell in the dry, wherever possible before post-tensioning. This is considered especially important in the case of larger ships, for reasons given in the section on procedural tests. However, if it is agreed that the soffit can be poured directly onto the dock bottom, arrangements to enable such an examination of the soffit to be effected on completion of construction should be established.

The site surveyor

It is evident that a prospective constructor of a concrete ship will have a long tradition of working with concrete or will have especially acquired an experienced staff and workforce to manage and execute the work. It cannot be presumed, however, that the construction company will automatically be responsible for the design. In many cases of civil concrete construction, the promoter of the enterprise appoints a consultant engineer to develop a design and the latter, because he bears prime responsibility for the design, monitors (*inter alia*) standards of construction through the medium of his resident engineer at the site.

There are thus two possible situations in which the site surveyor will find himself—the traditional one where the shipbuilder is responsible for design and construction and the less usual one in marine circumstances where the design and construction responsibilities are split.

It should be appreciated, however, that the site surveyor will be required by the committee to satisfy himself that the society's requirements have been met. Whilst he will expect to co-operate fully with the constructor's quality control team and with the resident engineer if there is one, he will be fundamentally independent of them.

Procedural tests

There has been considerable discussion in the industry on the possible construction methods for concrete ships. Whether conventional formwork or vertical or horizontal slip-forming techniques are adopted, whether the fore and after bodies of the ships are in steel or concrete, whether or not prefabrication methods are used, the essence of successful construction of a large hydrodynamically suitable concrete hull form is the production of relatively thin-walled, high strength (and thus highly reinforced and prestressed) sections requiring the tightest control of weight, concrete quality and dimensional tolerance.

The constructor should assure himself and all those concerned with quality that the intended construction procedures will lead to acceptable results, and to this end it is considered necessary to undertake procedural tests simulating the more difficult and untried pours and compaction techniques. The surveyor will participate in these tests. Typically they should include pouring and screeding a bottom and deck section and pouring a wall section selected to demonstrate that the material is sufficiently workable to be capable of being fully compacted into steel-congested areas without segregation, using the compaction methods intended for use on the ship itself. Similarly, grouting tests on cables should be carried out, simulating conditions on the ship, to ensure complete filling of the various duct configurations and terminations.

Other areas where compaction problems might exist, such as around fan anchorages associated with difficult vibrator access on account of formwork, may require to be investigated by means of procedural tests at the surveyor's discretion.

Forms of construction and associated problems of quality control

As stated earlier, the surveyor will, in general, require to examine the soffit of the bottom shell before post-tensioning (depending on the size and type of ship). This requirement is emphasized because, bearing in mind that this area will be highly congested with reinforcement, tendons and anchorages etc., it cannot be guaranteed that the soffit will be free of defects such as honeycombing and localized segregation without such inspection. The repair of such deficiencies after tensioning has been carried out presents obvious problems. Casting onto bottom formwork will have the added advantage of allowing a greater degree of freedom for the structure to move during tensioning and shrinkage, thereby promoting a greater confidence that the design stresses have been achieved.

Accordingly the site facilities should, in general, permit the bottom formwork to be adequately supported at a suitable height above the dock bottom, and the formwork itself should be of a high standard in terms of rigidity and watertightness.

One point worth considering here is the growing conviction among designers that to fully exploit the structural capability of concrete in marine applications, where axial and lateral loadings are to be accommodated, an arched slab construction is preferable to the flat slab alternative, despite the added frictional resistance of the former in service. In such circumstances the question of casting directly onto the dock bottom will not arise.

When considering the side shell construction, at first sight it may appear to some that casting a structure with three-dimensional curvature in concrete should be easier than producing the same shape in steel. In terms of prestressing problems and desirability of adopting advanced construction techniques, however, the difficulties of producing the fore and after body sections in concrete will have been one of the main preoccupations of the designer. Certainly, for the parallel middle body, vertical or horizontal slip-forming of the hull appears highly desirable both in terms of speed of construction and a reduced quantity of construction joints. The decision as to whether to adopt concrete or steel construction for the remainder of the hull is entirely his. It is not for a classification society to dictate in any way the basic structural philosophy of the hull, but it must be alive to the potential structural and practical difficulties involved in any option adopted by the designer.

The structural difficulties of either the composite structure (concrete mid-body and steel fore and after bodies) or the wholly prestressed concrete structure will largely have been resolved, at least in theory, at the design approval stage. However, it is to be expected that, in practical terms, there will be a considerable overspill of these problems into the production stage. The basic technical specification, which will be the prime document for the outport surveyor in his handling of the matter, will incorporate full details of the construction methods to be adopted. These will include any special shuttering techniques for highly shaped end sections, together with the associated procedural testing to be undertaken before construction itself commences.

The formwork required for decks, superstructures and associated work should be more conventional and thus present fewer problems of construction. The surveyor will no doubt concentrate his efforts on the areas of classical concern to the naval architect, namely the need for especially high standards of construction at deck to side shell connection, at discontinuities of strength, at scarfing in way of ends of superstructures of significant length, and at stress concentrations at the corners of hatchways and other deck openings. In this regard it is worthy of note that fatigue considerations are constantly in the minds of naval architects, especially when dealing with these areas, and the development of satisfactory detail design at such points in the structure should not be underestimated by the design team.

Formwork and placement of steel etc.

The economic viability of concrete ship construction requires that the tolerances for concrete cover are taken to the practical limit to attain minimum weight/strength ratios. Correspondingly, the accuracy of the placement of reinforcing steel and tendons must be of the highest order.

In common with the modern steel shipbuilding approach generally, the days when quality control was regarded by the production team as a constraint should be gone for ever. Speeds of production are, or should be, such that the quality control must be predominantly geared to prevent defects, not repair them.

With adequate planning and execution the surveyor should not have much comment to make when he carries out the final 'pre-pour' inspections. On these occasions he will be checking formwork, compliance with steelwork arrangement plans for scantlings and accuracy of placement, methods of reinforcement and tendon support and general cleanliness and surface preparation in way of the pour. The structural capability of both the formwork and any reshoring of the structure after its early removal is the responsibility of the relevant contractor. The surveyor will, however, be concerned in reviewing the dimensional checks made by the quality control team to establish that no significant deformations have taken place.

Mixing and application of the concrete

As mentioned earlier, the surveyor will have generally examined the site facilities for storage, handling and apportioning the various components of the mix and, of course, he will review mixing operations during construction to satisfy himself as to its uniformity in accordance with the agreed requirements.

In general it is expected that gravity pouring of the mix will be used exclusively during construction. Any special techniques for concrete placement such as application by spraying would be subject to full procedural tests before acceptance. The basic success of any concrete structure of the type under discussion depends on good workability and compaction control during casting. The vital objective, having attained an acceptable level of initial structural strength, is the preservation of that strength during the life of the ship in a salt water environment. Thus any voids, honeycombing or other significant defects of this type which could place the integrity of reinforcement or prestressing steel at risk from corrosion are inadmissible. Accordingly the surveyor will expect the main effort of the constructor's quality control team to be expended on the control of workability and compaction during concrete placement and he will participate fully with them. Subsequently, the control of the curing procedures and temperatures will be a major concern.

Construction joints

The procedure to be adopted for construction joints should be detailed in the technical specification. The degree of protection of steel reinforcement and tendon ducts against an ingress of water at the joints must be equivalent to that provided for the remaining structure. Thus the joints are to be tight across their full depth. The procedure should describe fully the intended surface preparations, epoxy coatings etc. and methods to achieve this end.

Post-tensioning operations

The designer will have precise ideas of what he wants to finally achieve by prestressing the structure and he may devise perfectly adequate procedures by which his objectives can be attained in practice provided the production schedule is fully maintained. The taking up of prestress by the concrete is, of course, a time-dependent process and the surveyor will be closely concerned in reviewing the construction team's efforts in maintaining the approved schedule in terms of sequence, timing, degree and efficiency. He should bring any significant deviations from the schedule to the attention of the plan approval centre for comment. The sensitivity of the final stress pattern to such deviations is much dependent on the structural arrangement concerned, and this question is worthy of further study by both the designers and the classification societies.

Inherently, large ship structures are likely to incorporate long tendons which are seldom, if ever, straight throughout their lengths. Accordingly, friction effects of the tendons in their ducts should be considered. For this reason, and also to give a clear picture of how the general prestress is being taken up by the concrete, the surveyor will examine the constructor's records of strain measurements along the lengths of the tendons. Each tendon should be identifiable with its individual record of prestressing operation (age of concrete at initial prestress, dates of prestressing, applied loadings as indicated at regularly calibrated hydraulic jacks, tendon extension expressed in terms of percentage of theoretical extensions, associated concrete compressive strains along its length and, finally, volumetric recordings of the amount of injected grout).

In general, any tendon wires which break during the prestressing operations are to be replaced.

Acceptability of the finished structure

The surveyor must decide in each particular case whether an imperfection must be repaired or not. This is because a criterion of acceptance in one location may not be valid for another. Not only will some parts of the structure be more structurally critical than others, but service conditions will vary, from zones which are either continuously submerged, intermittently submerged or salt-water sprayed, to the localities which are fully protected such as internal topside structure in the engine room. Moreover, due to the stress pattern across their thickness, cracked slabs which are predominantly loaded by local bending forces are not so susceptible to deep penetration by water as those predominantly loaded by axial forces, be they tensile or alternating tensile and compressive.

At one extreme it could be argued that unrepaired cracks and honeycombing etc., however minor, are inadmissible, whilst at the other extreme a case could be put forward for a fairly tolerant criterion such as accepting cracks less than 0·1 mm wide and a limited extent of surface honeycombing. In the latter case, however, judgement would have to be exercised as to the possibility of internal honeycombing being present on account of the same cause which promoted the surface imperfection. (A possible method of identifying honeycombed areas is discussed in the section on non-destructive testing.)

The alternative to this approach is to adopt a severe acceptance criterion and apply it to the entire structure.

Structural testing at site

Apart from the method of non-destructive testing, the general scope for site testing is naturally more limited than for steel ship construction, and it is for this reason that the procedural testing is considered so important. Nevertheless, systematic core sampling to an agreed and limited extent may be required to confirm that the target characteristic concrete strengths have been achieved. Moreover, carefully controlled shallow coring-out in areas presenting difficult constructional problems, such as under ducts in congested areas, can reveal the standard of compaction achieved. The use of radiography in such areas to indicate internal voids merits consideration.

However, the principal form of testing at site will be the hydraulic testing of watertight boundary structure. This is aimed at verifying both the structural capability to withstand hydrostatic pressures slightly greater than should normally occur in service and the integrity of the concrete as an impermeable barrier. The schedule of hydraulic testing should be specified at the outset and ideally should be arranged to simulate maximum global still-water bending moments and shear forces.

Since these tests are usually carried out near completion of the ship, special mention must be made in regard to proving the integrity of construction joints. These joints should be tested at a sufficiently early stage to enable any necessary repairs to be effectively carried out. Accordingly, local hydraulic testing of the joints by specially prepared pressure or vacuum jackets should be detailed in the technical specification. In the absence of such preliminary tests, the specification should indicate the procedures by which repairs may be effectively carried out during the later stages of construction.

PERIODIC MAINTENANCE SURVEY OF THE HULL

The periodic survey requirement for prestressed concrete ships will, in general, follow that laid down for steel ships. However, as already stated, the Committee of Lloyd's Register will assign a class notation of 'experimental' to such ships where it considers this appropriate. In such cases the periodical survey schedule will be augmented by additional examinations until this notation is withdrawn and the ship reverts to a normal class status.

For steel ships there are simple and effective means available to determine the deterioration in hull strength and integrity due to wastage and wear and tear, comprising visual inspection, pressure testing and the thickness determination of structural members. The nature of similarly effective inspection techniques for prestressed concrete ships must be established and the Society's approach is now outlined.

Annual surveys

Annual surveys are carried out afloat and the extent of the examination of prestressed hulls will be the same as that required by the rules for steel ships with the exception that for ships with an experimental notation an internal examination of cargo and machinery spaces will be required at each survey. The method of survey will be similar to that described for docking surveys.

Docking surveys

Ships with the experimental class notation will be required to dry-dock after six months' service. Thereafter, provided the findings at this first docking are satisfactory, the normal cycle of dry-docking surveys at two-yearly intervals will be required.

The docking survey is, of course, intended to reveal underwater damage in addition to the general condition of the hull and appendages. It is important that any damage to a concrete hull is repaired before more serious loss of strength is caused by corrosion of the structural steel. This is the main reason that dispensation from dry-docking between special surveys is not automatically given. Any request to dispense with such a docking survey on the basis of an underwater examination conducted in a similar fashion to that indicated in chapter R(K) of the rules for steel ships will, however, be considered. At both the annual and docking examinations the surveyor will be relying predominantly on visual inspection to detect signs of incipient corrosion of reinforcement such as spalling, cracking of the concrete associated with corrosion of tendons etc. and general surface defects attributable to wear and tear. Accordingly the surfaces should be reasonably clean, and a ready means of enabling close examination of all relevant areas should be at hand.

Should such visual inspection reveal significant defects requiring further investigation, the additional survey procedures mentioned in the following section on special surveys can be adopted as necessary.

Special surveys

Special surveys, which include a docking survey, are thorough examinations which vary in extent with the age of the ship as defined in the rules for steel ships. Accordingly the surfaces to be inspected are to be thoroughly cleaned. Selected structural tanks of the ship will be pressure-tested to an extent depending on her type and age.

What has been said in relation to visual examination during the intermediate surveys equally applies to the special survey, but a greater degree of survey preparation and a proportionately increased expenditure of time by the surveyor will be involved. Complete drying-out of surfaces to be examined during pressure-testing will be required. Normally these tests will be hydrostatic, but consideration will be given to some of the tanks being tested with air pressure or with vacuum techniques. The foregoing requirements are, of course, simple adaptations of the requirements for steel ships but special consideration must be given to methods of determining any deterioration of steel reinforcement or prestressing tendons.

Investigation of suspected deterioration

Spalled concrete shall be cut back to the affected reinforcement in order to ascertain the nature and extent of deterioration. Examination of the 'as fitted' drawings, in conjunction with the use of a cover-meter, will enable the distribution of reinforcement and the proximity of prestressing tendon ducts to the affected area to be ascertained. In the case where such proximity does exist, careful coring-out back to the external surface of the duct is to be effected to establish its condition. (Metallic ducts will give a better indication of possible tendon corrosion than will non-metallic varieties.) Should signs of duct corrosion be apparent, the tendons themselves must be carefully exposed to ascertain their condition also.

Similarly, where surface cracks give rise to concern for the integrity of the tendons or the reinforcing steel or both, the surveyor may require coring-out to establish the subsurface conditions.

Routine assessment of the condition of reinforcement tendons and associated steelwork

The extent of such assessment required at each special survey will depend on the age of the ship and whether or not the ship has the experimental notation. At the first special survey, in the absence of any surface manifestations of internal deterioration, a selected number of tendon anchorage end plates would be opened up for examination and reinstatement under survey. Their position and condition will be recorded by the society for the information of the surveyors at subsequent surveys. Naturally any deterioration discovered would prompt more extensive opening up at the discretion of the surveyor. At subsequent surveys other anchorages would be similarly examined in a cyclic manner.

At the third special survey of a ship with normal class status it will be necessary to verify and record the condition of principal structural elements at sections through the ship which typify its primary strength. It is to be expected that more refined methods of effecting this will be available by the time such surveys become due. Currently coring back to the tendon duct appears the only reliable method. If the steel duct surfaces show no signs of corrosion, it may be assumed that there is no diminution in strength, provided the history of anchorage condition is good. Should signs of duct corrosion be apparent, a representative selection of tendon wires are to be exposed and concrete cores tested to quantify the residual strength of the section.

Additionally, at this survey, selected positions on the external shell along the length of the ship between the light and loaded waterlines will be investigated by small-diameter coring to expose the reinforcement to verify its general condition. Other vulnerable parts such as the internal surface of the shell in way of hold bilges may also require routine investigation.

NON-DESTRUCTIVE TESTING DURING CONSTRUCTION AND MAINTENANCE SURVEYS

Up to this point, in the context of quality control and investigative surveys respectively, only direct material property testing and visual examination of embedded steelwork have been advocated, apart from passing references to the possible use of spot radiography to check for large voids in and beneath tendon ducts and to the routine use of the cover-meter.

This is because, on the one hand, the most promising alternative for measuring the characteristic strength of concrete during and after its hydration, namely ultrasonic pulse velocity testing, is somewhat comparative and is considered not yet to have reached the stage where the measurements can be authoritatively interpreted by any but the most specialized operators. On the other hand, no non-destructive methods appear to have reached the stage of development to measure the condition of embedded steelwork to the required accuracy of say 2% on the sectional area.

It is considered, however, that the ultrasonic pulse velocity techniques, properly used, are ideally suited to give an early general indication of poor concrete both in terms of its strength and homogeneity. Thus they should be valuable assets, along with the cover-meter, in the surveyor's endeavours of verifying that the careful controls prior to and during pouring have been successful.

Moreover, the ultrasonic technique also appears promising both as a means of detecting zones of deterioration in concrete strength due to fire, frost or chemical action, and as a method of locating gross voids, cracks or other imperfections.

It is evident that great scope exists for the development of methods to accurately verify both the continuity of and deterioration of prestressing tendons throughout their life, either with or without access to the anchorage plates.

REPAIR OF CONCRETE STRUCTURES

Repairability

The repairability of a wasted or damaged ship is not a matter for the classification societies, whose function it is to approve the technical adequacy of the repair proposals and to ensure that the work is effectively carried out. The owner of a concrete ship may confidently expect a relatively long natural life for his vessel and he will, no doubt, look to its designer for information on the economic repairability of a hull damage.

There appears to be very little explicit information published as to how, for example, the bottom shell of a ship which is prestressed both longitudinally and transversely can, in the event of damage, be partially renewed with the final levels of prestress in both the affected area and the surrounding structure reaching, and not exceeding, the required range of values. No doubt these problems can be alleviated to some extent by providing sufficient prestressing tendon couplers in the original design, but it is to be hoped that this aspect will receive wider attention in the future.

Wastage and wear and tear repairs

In the event of significant wastage of structural steel, it is to be supposed that such corrosion will be caused by a deterioration in the general condition of the concrete, together with its protective properties. The extent of such deterioration will be established by the surveyor to enable him to make a judgement as to whether a renewal of the classification certificate for a further five years is warranted without recourse to structural repairs. In certain cases it is possible that he will need to seek advice from the plan approval office where a more analytical evaluation of the situation is possible.

Should the wastage be severe and widespread, the owner will have to consider the technical and economic viability of general renewal of the structure. In less extreme cases, proposals for additional prestressed reinforcing structure to be integrated with the existing hull girder to restore general strength will be considered and approved subject to the arrangements being adequate in design and execution.

Repairs to wasted reinforcement may be carried out by local renewal with adequate lapping of the bars, followed by overlaying with properly bonded non-shrink concrete or other approved structural material, provided the stresses which the repaired area (which will not be prestressed unless this can be arranged by selective ballasting of the ship) will be subject to in service are acceptable. Such proposals may require analytical confirmation as the field surveyor will not, in general, be in possession of the information required to enable an authoritative decision to be made on the spot.

Minor surface reinstatement can be carried out without such reservation either by epoxy patching using approved procedures or by fine concreting using non-shrink cement or admixtures with keyed joints to the satisfaction of the surveyor.

Damage repairs

General. Both the owner and the underwriter will normally expect damage repairs to be such as to place the structure in the same condition as existed prior to the casualty or to an acceptable equivalent condition. Minor surface damages can be dealt with as described above for wear and tear repairs.

Major damage repairs. The society will normally have to make a preliminary assessment of the extent of major damage and the residual strength and capability of the ship to enable her to proceed to a suitable shipyard. Professional divers' reports may be necessary in certain cases.

Temporary repairs may have to be made and the class status of the ship may have to be reviewed before such a voyage is made.

Proposals for the permanent repairs will define the extent of the cutting away of the damaged and surrounding structure necessary to permit full reinstatement of prestress to the affected area without disturbing the integrity of the remainder of the hull. Once these proposals have been agreed, the procedures will largely follow those laid down for new construction. It will be important, no matter how old the ship, that full design information is retained to enable such proposals to be formulated. It may be that special ballasting configurations can be devised to bring the affected zone into a state of near zero compression to facilitate the repair by enabling the new structure to take up the required prestress in the desired fashion. However, it must be borne in mind that damage repairs to the bottom using such a technique must be dealt with in dry conditions, and repairs near the neutral axis cannot be approached in this way.

Repair of cracks. There exists considerable specialist expertise for epoxy injection of cracked concrete and provided the resin mixes are carefully matched to the job in hand in terms of viscosity, setting times and bonding ability in damp conditions, and that the injection procedures are adequate, such repair of benign cracks is acceptable to the Society. Measurable cracks existing in a prestressed area under conditions of normal loading are indicative of damage to the tendons, and such damage will require proper investigation before the crack itself can be dealt with.

CONCLUSION

The historical function of Lloyd's Register of Shipping has been to set internationally acceptable standards of ship construction and survey and to continuously update these standards both on the basis of the service performance as reported by its field surveyors during the life of classed ships, and on the basis of the advances made in technologies of materials, load assessment, analytical procedures for determining structural response to these loads, and construction methods.

The renewed interest in the possibilities of prestressed concrete for the construction of large specialized ships has prompted a concentrated effort on the Society's part to consolidate on the available experience of the service histories of the smaller reinforced concrete vessels, and to project its traditional expertise on ships' structures and analytical techniques into this new field of endeavour, to provide all those concerned with a sound basis on which to proceed.

This paper has not touched upon all aspects of quality control and maintenance. For example, the topic of the cryogenic application of concrete in this context has not been discussed. Other questions such as the procedure for large complex repairs have been raised and left open for further thought and discussion by all concerned. However, it is hoped that the Society's approach to these matters has been clarified. Much will be required of the field surveyor in terms of competence to deal with such structures and of constructive effort to assist both builders and repairers and their quality control teams. His major concern will be to promote the assurance of quality and to think predominantly of prevention rather than cure. Above all, the Society looks forward to full participation with all those interested in the development of concrete vessels. It does so with full consciousness of its obligations to encourage the efforts of the innovative engineer and to provide owners, shippers and crew with a sound and safe ship.

REFERENCE

1. ALGATE C. F. The role of classification. *Concrete afloat.* Thomas Telford Ltd, London, 1977, 43–50.

BIBLIOGRAPHY

AMERICAN CONCRETE INSTITUTE. *ACI manual of concrete inspection.* American Concrete Institute, Detroit. Sixth edn. 1975.

HARRIS A. Characteristics, factors and design criteria. *Proceedings of the conference on concrete ships and floating structures.* University Extension, University of California, Berkeley, 1976.

LEONHARDT F. Prestressed concrete ships by slipform construction. *Proceedings of the conference on concrete ships and floating structures.* University Extension, University of California, Berkeley, 1976.

MULLER J. Structural considerations and configurations. *Proceedings of the conference on concrete ships and floating structures.* University Extension, University of California, Berkeley, 1976.

TOMSETT H. N. Site testing of concrete. *Br. J. Non-Destr. Test.,* 1976, May, 82–87.

Concrete afloat. Thomas Telford Ltd, London, 1977, 147–158

The paper discusses the history of concrete vessels, proposed present-day construction methods, the selection of suitable construction sites, general suitability and weight considerations of concrete vessels compared with conventional steel vessels, special applications of concrete, especially in regard to cryogenic cargoes, future possibilities with concrete as the building medium and the present state of the art.

Practical design and construction considerations on the use of prestressed concrete in ship construction

S. I. G. TOSSWILL, MRINA*

The use of concrete as a construction material for the hulls of seagoing vessels and structures has on most occasions in the past been brought about by war situations and the consequential need for savings in the use of steel. World Wars I and II were both periods during which concentrated efforts were made to use reinforced concrete as a shipbuilding construction material, the period between the two wars seeing little use of concrete as there was once again a plentiful supply of steel.

During the 1914–18 period many concrete ships and barges were constructed both in the UK and the USA. The best known UK vessel of this era was probably the SS *Armistice* which, launched in 1918, had a dead-weight capacity in the order of 1100 tons and traded successfully for 25 years or more. At about the same time an all-concrete US ship was delivered, somewhat larger than the *Armistice*, with a dead-weight capacity in the order of 4000 tons. Both these vessels had notable careers and between them carried out many long ocean voyages. It could be said that they were the pioneers in concrete shipbuilding as far as ocean-going vessels are concerned. Many concrete ships continued to operate after World War I through the 1920s until, due to the slump and trade recessions, they were finally withdrawn from service, many of them finishing their lives as a mute testimony to their durability and strength as breakwaters in various ports throughout the world, some still remaining to this day. Indeed, the removal of their remains has sometimes proved very difficult and a serious problem.

Between the two World Wars an inevitable recession in the use of concrete as a construction material for seagoing ships took place with the return of plentiful supplies of steel. A few hulls, however, were built in the early 1920s in France, these vessels having a dead weight of about 1300 tons.

* Naval Architect, Burness, Corlett and Partners Ltd, Basingstoke.

The return of concrete as a shipbuilding material did not come about again in any major way until World War II. During the early 1940s, the USA took the lead and contracts were awarded for the building of a considerable number of self-propelled concrete ships. Although in the UK during the period of World War II only two self-propelled concrete ships were built, a most important UK contribution towards introducing floating concrete hardware was the construction of the Mulberry harbours, the large concrete structures built to form breakwaters and other specialized harbour equipment. These units were suitable for towing across the English Channel and formed sheltered water areas for the landing of troops and equipment.

The construction methods of the early 1940s were basically no different from those of the 1914–18 era but the vessels were much stronger, due mainly to better concrete mixing and better cement. All the vessels were still of reinforced as distinct from prestressed concrete: the latter would now, of course, be the choice, quality also being raised with modern technology and quality control.

CONSTRUCTION METHODS

It is certain that, in the light of modern technology, concrete ships built today must be constructed using prestressed rather than reinforced concrete. Prestressing is now a fully accepted construction method in the civil engineering world at large and has been used extensively also in offshore installations and many floating and static structures operating in a marine environment.

Prestressing techniques entail the improvement of the inherently low tensile strength of concrete, by using steel tendons under tension. Thus a permanent initial compressive stress is induced in the concrete and the whole cross-section is utilized to resist subsequently applied tensile stresses. These, as an absolute maximum, could go up to that of the precompression applied in the tendons, or to the level of stress necessary to avoid concrete tension if used for ships' hulls when operating under service conditions.

The assembly techniques used for the building of a concrete ship should certainly be on a prefabricated system whereby units are produced under factory conditions adjacent to the erection site. The prefabricated units are each individually prestressed and then post-tensioned one to another at the erection site in a longitudinal and transverse direction. As naval architects, Burness, Corlett and Partners have worked with various civil engineering bodies on the development of concrete structures for shipbuilding purposes, advising mainly on the maritime aspects of design, and the international regulations to be complied with and necessary when designing ships, whatever their construction method or material. Much work was done and preliminary investigations were carried out with F. H. Turner in the development of prestressed hulls for large cryogenic carriers (liquefied natural gas (LNG)). The same type of construction is, however, applicable to any concrete floating structure and is basically an application of the prestressed–precast construction method already referred to.

It is considered that vessels to carry LNG are probably one of the best applications for concrete as a construction material in the light of the advantages to this type of carrier with cargo stowing at around 85 ft^3 to the ton. The reason is, of course, that the low specific gravity of the cargo makes the vessel capacity limited, while the high structure weight relative to steel is no disadvantage and helps, when light, as the use of water ballast is minimized.

A carrier with a cargo capacity of 15 000 tons (Figs 1 and 2) was selected as basis ship for fairly detailed study, and approximate data were derived for carriers with capacities of 5000, 10 000, 20 000 and 25 000 tons capacity. Table 1 shows the outline dimensions of the vessels considered.

The 5000 and 10 000 dead-weight sizes were discarded early in the investigation when it became apparent that economics strongly favoured the larger sizes.

Design is based on a honeycomb cellular system with what amounts to virtually inner and outer hulls. The spaces between the two hulls form a cellular void between the sea at the outside of the outer hull and the cargo

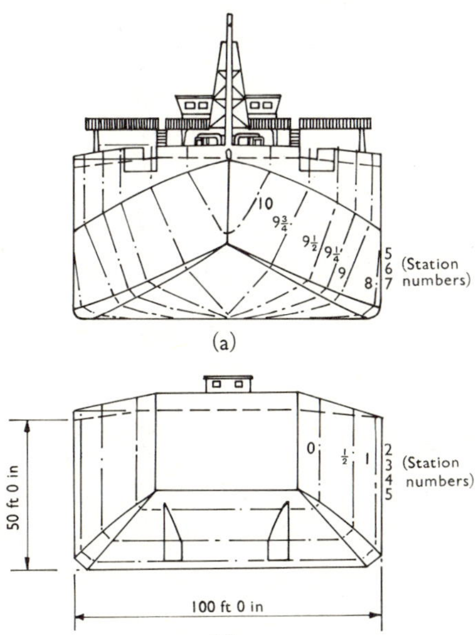

Fig. 1. Barge of 15 000 tons nominal dead weight: (a) bow elevation; (b) stern elevation

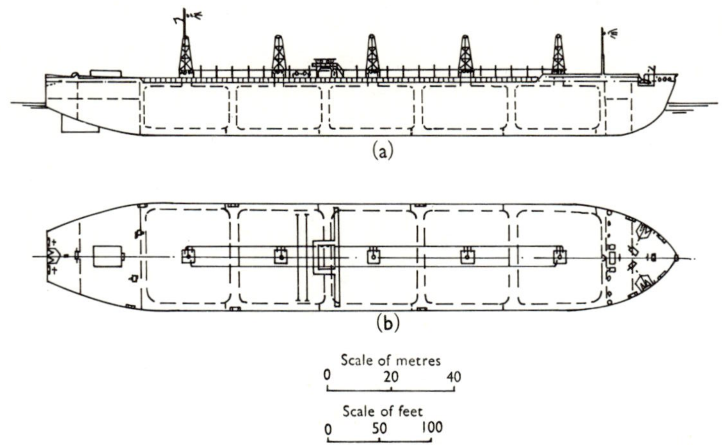

Fig.2. Barge of 15 000 tons nominal dead weight: general arrangement; (a) profile; (b) main deck

Table 1

Cargo dead weight, tons	Length, m	Breadth, m	Depth, m	Draught, m	Light ship, tons	Other dead weight and margin, tons	Displacement loaded, tons
5 000	134	22·6	11·3	7·9	15 200	600	20 800
10 000	165	27·5	13·7	8·5	22 600	900	33 500
15 000	183	30·5	15·3	9·1	28 000	1400	44 400
20 000	197	33·0	16·5	9·7	33 000	2000	55 000
25 000	213	35·5	17·7	9·9	37 700	2600	65 300

tank at the inside of the inner hull. The basis vessel, the 15 000 ton deadweight carrier, has five cargo compartments, each having double transverse bulkheads at its ends. Thus each tank is contained within its own void space which in turn is charged with inert gas. The inner surface of the inner shell forms the cargo tanks, and to this surface the insulation and membrane lining are attached to maintain the liquid methane at about $-162°C$.

As designed, the carriers have a single flush deck with bulwarks forward and aft and a monitoring/conning bridge near amidships (this bridge being for conning the carrier when entering or leaving harbour and for housing instruments and cargo monitoring equipment when loading and discharging). A deckhouse aft covers the machinery and service space and also houses equipment such as fans, ventilation equipment and such items associated with this type of vessel.

As we have considerable experience in the design of high efficiency single-curvature hull forms and as the shaping of the concrete prefabricated sections will thereby be much simplified—only single curvature is used—this type of shape was adopted. The whole family are of ship-shaped, full-bodied form, with single curvature and double chine section, transom stern and raked bow. The basic concept was for towed barges and in that configuration twin skegs are fitted in a good position decided by the tank testing; they give, in the normal way, good directional stability and minimum drag under service conditions. However, where the barge is to be pushed, such skegs are not necessary but a single centre-line skeg is required for docking and for directional stability. With care such hull shapes can be designed to give a performance fully equal to that of the most complex double-curvature shape. This is a factor of considerable importance for concrete ships and one of which the concrete specialist may not be aware.

The structural methods finally used were based on calculated longitudinal bending moments and shearing forces which were carried out for hog and sag conditions of the ship girder on the basis of a wave height of one twentieth of the length of the vessel; thence, forces in the keel and decks could be determined. From these forces the constant hog–sag reversals of the girder can be catered for, with the steel tendons providing tension forces and the concrete contributing its strength during the compressive section of the reversal cycle. The concrete also contributes to resistance in shear and transverse moments and provides a watertight skin to keep the steel prestressing and post-tensioning materials in their design positions.

In the resulting structure using prestressed concrete, no tensile stress occurs in the concrete below the waterline, whilst a concrete tensile stress of 35 kg cm^2 arises at deck level as a peak stress of rare occurrence. The inner and outer faces of the hull are designed as panels to resist local water and LNG pressures as appropriate. While worked in some detail, the design has still to be finally optimized. It seems likely that some economy could be obtained by using thinner bulkheads in conjunction with a more detailed analysis of the prestressing tendons.

The containment of the cargo within the concrete tanks is intended to be of the membrane type, where the inner surface (the membrane) is of welded aluminium or stainless steel. Behind the membrane a load-spreading layer of lightweight/insulating concrete transmits the cargo pressure through the insulation to the inner face of the hull. The insulation consists of three layers of foam glass with a total thickness of 15 in. (20 in. at the bulkheads). The inner face of the concrete is to be coated with reinforced glass fibre to provide a secondary barrier for the LNG.

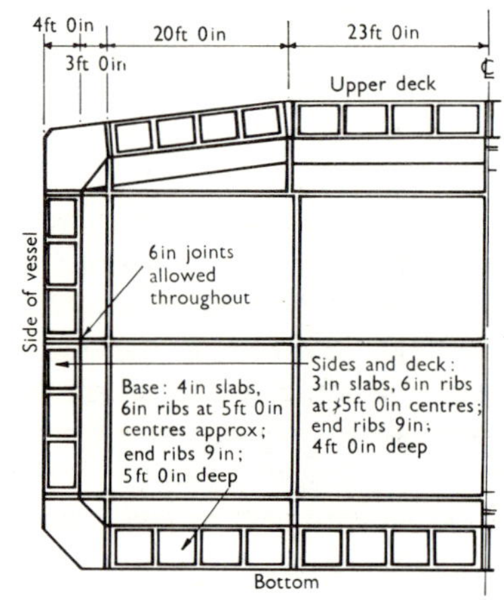

Fig. 3. Section through carrier

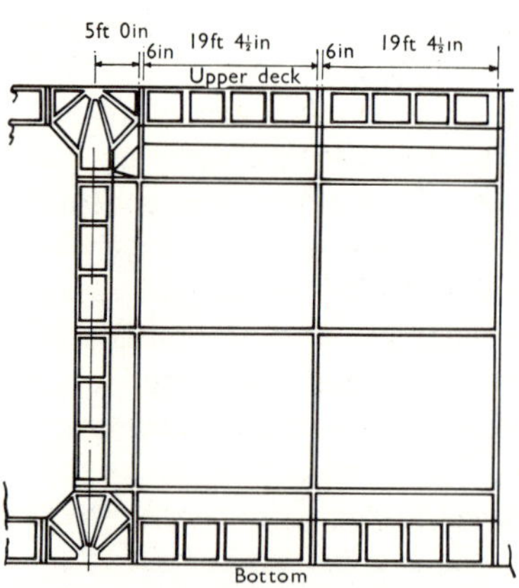

Fig. 4. Sectional profile through tank and transverse tank boundary bulkhead

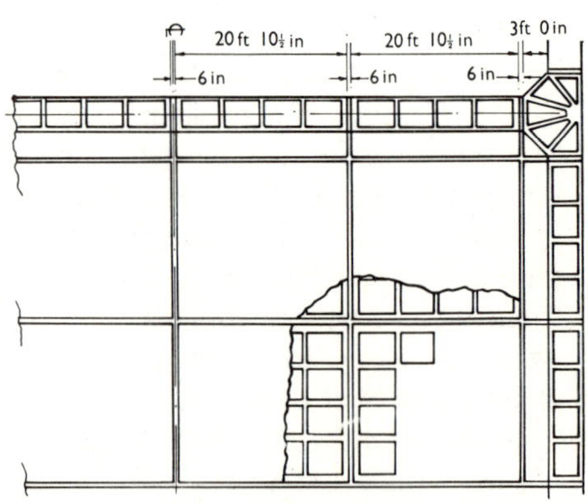

Fig. 5. Sectional plan through tank and transverse tank boundary bulkhead

An attractive possibility is to use a prestressed concrete liner in place of the aluminium and stainless steel membranes at present used, and a certain amount of experimental work in this connection has been and is being carried out.

The prestressing of concrete is a specialized technique; as its objectives and methods are not widely known outside the civil engineering industry, it may be useful to expand on the methods of prestressing which it is anticipated would be used in the construction of the carrier investigated. It is necessary to use high tensile cryogenic steel to induce the precompression into the concrete and to allow a margin to take account of loss of prestress which occurs during the weeks after stressing is completed. There are two standard methods of prestressing: the first—pre-tensioning—involves the initial tensioning of the steel before the concrete is cast. Once the concrete is cast around the steel and hardened, the steel applies the prestress to the concrete by bond and the temporary anchors holding the steel can be released. The second method—post-tensioning—involves cables or tendons placed inside ducts in the concrete which has been cast already. After the concrete has hardened, the cables (which are still free to move) are tensioned by means of jacks bearing against the concrete. Mechanical anchors are attached (there are different types of these), the ducts grouted up and the jacks released. The compression is applied to the concrete through the mechanical anchors and the grouting. The duct grouting also prevents corrosion.

These methods when applied to the proposed carrier result in a judicious combination of both methods. The separate panels, each about 6 m square, include some pre-tensioned steel, while post-tensioning is used to give the necessary strength across the in situ joints between the panels and to provide sufficient overall prestress to meet the worst anticipated service conditions.

The construction method described and illustrated is what we visualize would be used today in the construction of fairly large prestressed concrete vessels. Although the description has been mainly in the context of an LNG carrier, it is considered that basically the same methods of construction would be used for any type of floating structure, propelled or otherwise (Figs 3–5).

CONSTRUCTION SITES

One of the real advantages which the construction of a vessel in prestressed concrete affords is the relative ease with which it can be constructed without special shipyard facilities and without the need for highly developed technology in the country of construction or a vast reservoir of skilled labour and expensive plant. Furthermore, most of the materials are likely to be obtainable locally, saving both foreign currency and transport.

The construction site can be on any protected shoreline with a reasonable beach which is not too remote from main road access, from which feeder roads to the construction site keep the fabrication plant fed with a good supply of local aggregate.

The construction site basically consists of a construction dock and a precasting plant for the fabrication of units. The construction dock is formed by excavation, whereby the sides are graded to a stable slope, and the dock entrance is closed by sheet piling. The precasting plant for the fabrication of units is set up at the head of the dock, the concrete units being cast and prestressed using the pre-tensioning method as previously described. When the castings have hardened sufficiently, they are removed from the casting beds and transported to a stockyard or directly to their final position within the dock. The erection area could be serviced by a gantry type crane which could transfer precast sections from the casting beds to the mobile lifting frames, two of which would be erected—one port and one starboard of the building berth (Fig. 6).

When the first vessel is complete and erected in the dock, the sheet piling at the dock entrance is removed and the hull floated out, after which the vessel proceeds to the quay or berth formed by the material excavated for the dock. The sheet piling is then re-driven, the dock is pumped dry and the construction of a second vessel is commenced. In

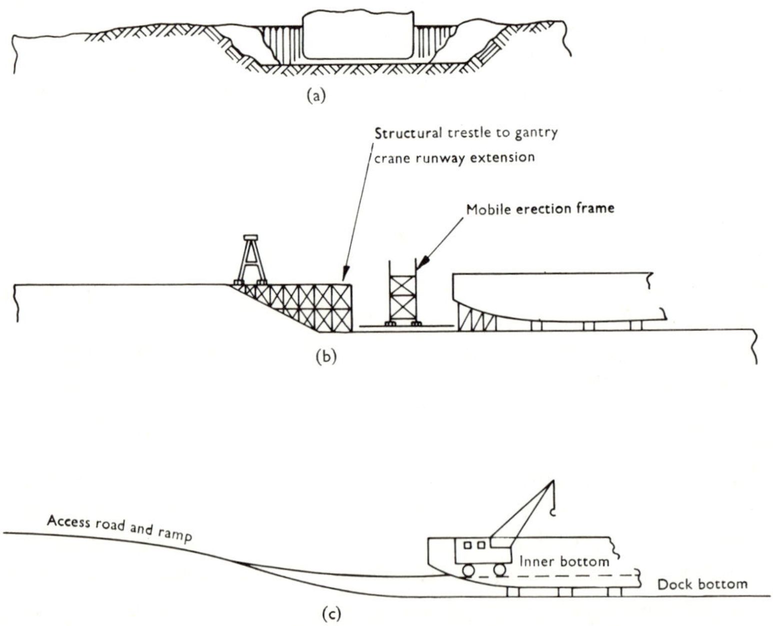

Fig. 6. Proposed construction dock arrangement: (a) cross section; (b) longitudinal section, trestle gantry access; (c) longitudinal section, centreline ramp access

this way it is estimated that, once the dock had been set up and production and supply lines streamlined, it would be possible to produce a 600 ft carrier in about every 10 months.

An alternative method of building is shown in part (c) of Fig. 6. This again entails the excavation of a construction dock, but instead of the trestle gantry and erection frame, a ramp with a suitable declivity is cut at the shore end of the dock to coincide at the dock bottom with the inner bottom height when erected on the keel block. It is thus possible with suitable mobile cranes to start erecting the bottom structure with the mobile cranes starting at the ramp end (shore end) and using the double bottom structure along the centre as an erection roadway, with the sides, decks and bulkheads built up in the reverse direction, from seaward to shore end. On completion the procedure is similar to that in the previous method, entailing the floating out of the vessel, re-driving of the sheet piling and pumping-out of the dock ready for another vessel.

Depending on the coastline selected as a construction site, there will be many variations on the proposed methods and layout as described above, but it can be said fairly certainly that the capital overheads involved in setting up a construction complex are relatively modest if a number of vessels are to be built.

WEIGHT AND GENERAL SUITABILITY

It can be argued that the biggest reason for not using concrete as a construction material for ships is the penalty that an owner has to suffer on weight. Due to the much larger light-ship weight of a concrete ship, a reduction in dead weight when compared with a steel vessel of the same dimensions is inevitable and leads to a correspondingly low dead weight/displacement ratio. The loss of dead-weight cargo carrying capacity with a concrete vessel is for general purposes a valid objection but has tended to lead to the belief that concrete as a material is unsuited to ship construction generally.

There are, however, several types of vessel for which concrete has many desirable advantages and is better adapted to the purpose than steel. Included in these types are pontoons, where the weight is relatively of little importance, small craft where the weight can be reduced by using the ferrocement technique, and all types of vessels which are capacity carriers as opposed to weight carriers.

One of the most suitable applications of concrete as a ship construction medium is in the carriage of a cryogenic material such as LNG which, besides its low density, is carried at very low temperatures. Concrete is one of the few materials capable of resisting conditions imposed by very low temperatures, and at the same time remaining acceptable as a structural material. It is possible to reduce substantially the expensive protection which is needed in steel ships to guard against the leakage of LNG at

−162°C which might come into direct contact with the steel structure, leading to material failure. The possibility of a failure of this type does not exist with a prestressed concrete hull, provided suitable reinforcement material is used in the containment area.

Two main considerations affected the general suitability of prestressed concrete as a construction material for the 15 000 ton LNG carrier investigated. First, the increase in weight over steel and the consequential addition to the light-ship weight was not significant when carrying a low density cargo such as LNG; second, it was found that the difference between light and loaded draught was quite moderate so that the addition of water ballast in the light condition was not necessary, either from the point of view of getting the vessel down in the water or for stability. This is not the case with a steel LNG carrier where considerable water ballast has to be carried both light and loaded.

Cryogenically it has been proved from various studies that prestressed concrete itself is a very suitable material for use at temperatures in the order of −162°C. In fact, at these low temperatures the tensile and compressive strength of the concrete increases, but care must be taken with the reinforcing material which should be of cryogenic nickel steel to cope with temperature creep through the concrete. Probably this would be necessary only in the inner concrete structure adjacent to low temperature cargoes or materials.

In the carrier proposal discussed, it was not intended that the prestressed concrete hull of the carrier would be subjected to cryogenic temperatures, but if an accident should occur the behaviour of a concrete hull in contact with LNG would be totally different from that of a steel hull and would sharply reduce the need for expensive protection against LNG leakage in the form of secondary barriers.

In addition, prestressed concrete would also be acceptable from the weight point of view for custom-built vessels for the carriage of cars. Concrete car decks could be utilized with conventional 'tween-deck hatch arrangements.

Generally, any vessel which is destined to carry cargoes, solid or liquid, stowing at a rate of 85 ft^3/ton or above, would be a possible contender for construction in concrete.

The carriage of containers is another possibly suitable application for concrete construction. The average gross weight for a loaded 20 ft standard container is in the order of 15 tons, so a stowage rate of 85 ft^3/ton is obtained, exclusive of the volume lost in the holds due to container guides and access facilities.

It is possible to enumerate many hypothetical possibilities for carriers to use concrete construction. There is no doubt that because of the twofold advantages mentioned earlier when carrying cryogenic cargoes, this application is very attractive. However, any capacity carrier can be a viable contender for construction in prestressed post-tensioned concrete.

Designing an LNG carrier on a single-skin basis, not cellular, was considered. It was decided that in view of collision or grounding risks the double-hull type structure was safer for the carriage of liquid cryogenic cargoes. A major consideration was the cover afforded against pollution and other hazards, especially within a discharging port complex. When the LNG carrier design which has been described was conceived in 1970, it was considered to represent a step forward in the use of marine concrete and concrete technology. It meant that concrete could be introduced to the marine field in a practical and economic manner. It is still believed that this is so. In spite of the seven years that have elapsed since conception, the design as proposed is still viable and in keeping with present-day technology and thinking. Natural development will obviously come to light should the designs be brought to a final development stage for construction.

FUTURE POSSIBILITIES AND SUMMARY

It cannot be said that naval architects in general are highly enthusiastic at the thought of prestressed concrete as an alternative to steel. In fact it must generally be accepted that it is the civil engineer who comes to the naval architect and says 'What about a concrete ship?', rather than the naval architect approaching the civil engineer.

There is still a need for much large-scale testing and experimental work to verify and optimize the make-up of concrete and its reinforcement, with an aim to producing possibly a lighter and stronger concrete mix suitable for marine purposes. Again, the best distribution of steel in the structural section is vital—after all, the reinforcement alone is 20–30% of the steel weight of an all-steel vessel.

Given a reduced light-ship weight, due either to improved concrete or an optimized midship section giving improved local and overall distribution of material, then the real advantages of concrete for ship construction purposes would become more evident. These are:

(*a*) ease of construction which lends itself well to a production line;
(*b*) reduction of construction costs compared with steel vessels;
(*c*) reduction in maintenance costs throughout the life of vessels;
(*d*) the special qualities of prestressed concrete when applied to cryogenic carriers.

The routine concrete ship has not yet become a reality, but more people are becoming conscious of concrete as a possible alternative ship construction material in the light of ever escalating steel and labour costs. The time is right for civil engineers and naval architects together to solve, in the light of modern technology, those problems which still exist and to minimize the disadvantages of the material.

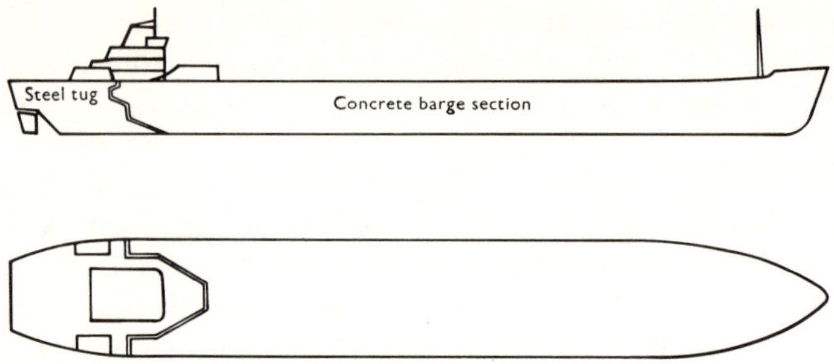

Fig. 7. Murvicker or similar tug–barge unit

There could well be a half-way house where vessels are built with conventional sterns containing the machinery, with prestressed concrete middle bodies. It would appear that the best application for prestressed concrete at present, when thinking in terms of seagoing ships, is that of the cryogenic carrier for the carriage of LNG. A marriage of the two with an autonomous steel propulsion unit is an attractive possibility.

The unit could be of the Murvicker or a similar type of integrated tug–barge unit. The result would be a steel tug unit containing propulsion and power units, and a prestressed concrete barge unit. A very flexible transportation system could be developed with three concrete barges and one tug unit, where the tug unit was kept permanently employed in moving cargo on the basis of one barge at sea and one barge at each port loading or discharging (Fig. 7).

Perhaps the hardest task that devotees to the idea of concrete ships have is to sell the idea to ship owners who, as always, have to be sure of the product they are buying and be certain that it will run reliably and on a profitable basis. It is always difficult to get new ideas off the ground. This can even be said about steel itself when first introduced to the shipbuilding world, while another obvious example was the advent of the container ship which initially was not very popular with ship owners until it was realized that it was a swift and efficient method of carrying cargo.

Concrete afloat. Thomas Telford Ltd, London, 1977, 159–165

Concrete is in many ways particularly favourable as a building material for gas ships. However, both gas tankers and steel shipyard capacity are presently in substantial surplus. The authors review the market for gas tankers and the situation of shipyards, and conclude that it will be the mid-1980s before demand for new gas tankers combined with other shipping will place a strain upon the capacity of conventional steel yards. Efforts to market a concrete vessel will face very stiff competition until that time. If concrete gas ships are to be built for commercial purposes, the authors foresee the following minimum preconditions: market conditions must improve substantially; a prototype must be built and operated to remove all questions about technical feasibility; and concrete shipyards should be created which meet the contractual needs of shipowners, including a substantial cost saving v. steel construction.

The market for concrete gas carriers

N. E. HAY*
C. JANSKY†
F. MASCARO‡

This paper is based on the experiences of the authors in gas shipping and in six years of developing and marketing concrete marine structures for the transportation and storage of liquefied flammable gases. The first unit to reach the construction state, the 60 000 cbm§ liquefied petroleum gas (LPG) storage barge *Ardjuna Sakti*, was delivered to Atlantic Richfield of Indonesia in June.

For purposes of this discussion we take the term 'gas carriers' to mean primarily self-propelled ships rather than barges, because these constitute the potentially larger and more generally interesting market. This will also considerably simplify the analysis. However, many of the comments also apply to barges.

The design and feasibility of concrete liquefied gas ships has been a subject of attention for nearly a decade. Recent papers have set forth at length the advantages of concrete construction for cryogenic gas carriers, and we make no further effort here to prove that concrete can be structurally equivalent to steel as a shipbuilding material. Suffice it to say that concrete is in many way particularly favourable for cryogenic gas shipping. The favourable properties include the following: favourable mode of failure under accident conditions; superior strength at cryogenic temperatures; resistance to fire and explosion; favourable insulating characteristic; high resistance to fatigue; lower maintenance costs; and, hopefully, lower construction costs.

* Vice President, Trans-Energy International, Inc., Hamilton, Bermuda.
† Vice President, Trans-Energy USA, Inc., Houston, Texas.
‡ Managing Director, Trans-Energy Services Ltd, London.
§ Cubic metre.

GAS TANKER SUPPLY AND DEMAND PROJECTION

Figure 1 shows the supply and demand of liquefied natural gas (LNG) (methane) tankers of over 5000 cbm capacity for 1965–85, with the future projections based on the existing fleet plus firm new vessel orders and existing LNG production plus firm new production projects. For convenience, fleet capacity and requirement are expressed in terms of a unit called a 125 000 cbm equivalent, which is analogous to the T-2 equivalent used in the oil business. This unit expresses laydown capacity or requirement in terms of the standard-size vessel in this trade today.

As Fig. 1 illustrates, the number of vessels in the LNG fleet will more than double between 1975 and 1980 from 29 to 67 vessels, resulting in a more than fivefold increase in laydown capacity. The projects for which most of these vessels were originally intended have been delayed to 1980 and beyond, largely because of political and regulatory factors, with the result that the present surplus will grow dramatically to over half the LNG fleet by 1980. Even if we add those projects which are possible by 1980 but not firm so far as the timing, the fleet will remain surplus at that date.

Beyond 1980 there is a wide range of uncertainty, depending principally on the world economy and import policy decisions which are taken in the USA and Japan. If all of the announced projects were to materialize, a substantial requirement for new vessels would occur in the early 1980s, reaching some 88 vessels of the 125 000 cbm size or their equivalent in terms of laydown capacity by 1985.

What Fig. 1 shows is that there will almost certainly be no market for the delivery of new LNG vessels before the early 1980s, but that there will possibly be a requirement for new vessels by 1985 which could reasonably be for another 20–30 ships.

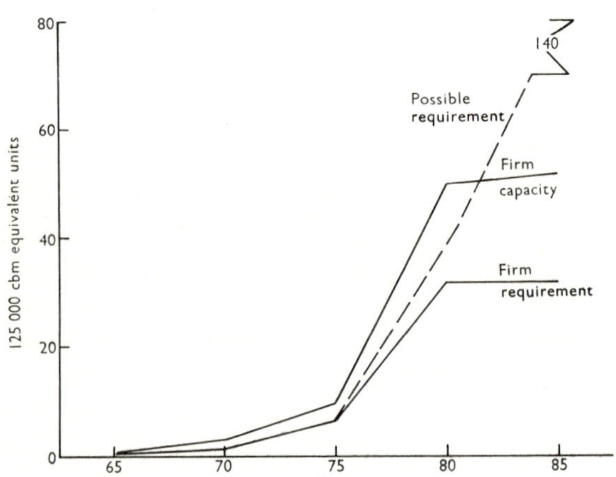

Fig. 1. World LNG tanker fleet: transportation capacity and requirement (vessels exceeding 5000 cbm capacity)

In contrast, for LPG (i.e., propane, butane, anhydrous ammonia, and petrochemical gases such as VCM and butadiene) there is a highly developed commercial trade. The LPG tankers are seldom built for a particular route, moving a given contracted quantity of gas, with the result that forecasting for LPG tankers is subject to an even higher degree of uncertainty than for LNG tankers.

In view of these uncertainties we have prepared a base case reflecting the events which we expect to occur, and two sensitivity cases. The resulting very broad range of possibilities unfortunately reflects the actual degree of uncertainty.

These cases are shown in Fig. 2. As in the LNG tanker analysis, vessel availability and requirement is expressed in terms of the laydown capacity of a standard-sized vessel for the trade, in this case a 30 000 cbm vessel. The base case future projection reflects only firm new-building orders and firm LPG production projects. We have also added a reasonable estimate of the transportation requirement for ammonia and petrochemicals, which is not a sensitive variable.

As in the case of LNG tankers the outcome of the analysis of LPG tanker supply and demand is largely determined by US and Japanese im-

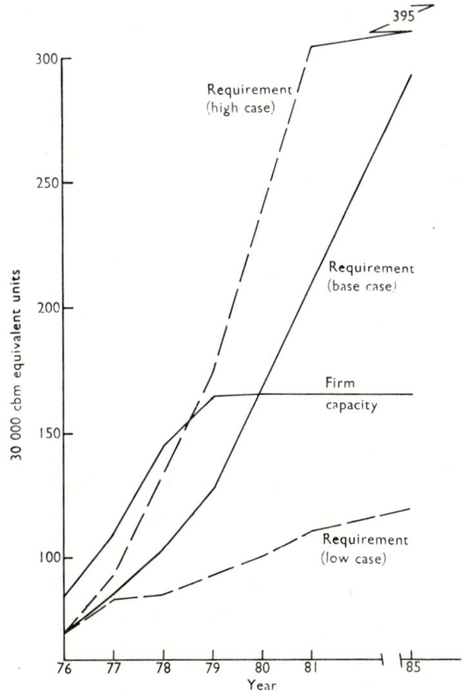

Fig. 2. World LPG tanker fleet: transportation capacity and requirement (vessels exceeding 10 000 cbm capacity)

port policy decisions which have yet to be taken, and to a smaller degree by the timing of new export projects. In the base case we have increased projected Japanese imports modestly from those of the April 1976 MITI forecast, rising to 15 million tons by 1981 instead of the 12·5 million tons projected by MITI. We have supplied all of the traditional import markets, with US imports constrained to the remaining export production. On this basis US imports are projected to rise to 9·4 million tons by 1980, amounting to 17% of domestic demand. (Further discussion of the US market is given by Hay.[1])

In this base or expected case the current surplus of LPG tankers will continue to 1980, reaching a peak of the equivalent of some 41 vessels of 30 000 cbm. This surplus is largely in the very small and very large vessels, and it is possible that demand for new buildings in the middle range will occur before 1980, especially since these are the logical sizes to serve existing US facilities. Beyond 1980 the base case projects a demand for new LPG tankers rising to the equivalent of 129 tankers of 30 000 cbm by 1985. This upside potential, based on firm production projects, is substantial.

However, we must also look at the sensitivity cases. The requirement for new tankers by 1985 in the base case does reflect firm production projects, but it rests on the assumption that the US and Japanese markets will absorb this new production. In the low-case assumptions US imports remain at about the 1976 import level, and Japanese imports are according to the MITI forecast. In this case a very substantial LPG vessel surplus of all sizes will continue until 1985. There is a serious risk of such a continuing surplus if US policy and US facilities do not permit a severalfold increase in LPG imports by the 1980–81 period.

In the high-case assumptions Japanese consumption is again constrained to the MITI forecast levels, in this case freeing additional quantities for the longer trip to the USA, and additional Persian Gulf production comes on to permit US imports to rise to 13 million tons by 1981. In this case the entire fleet would come up short by 1979, with the shortfall rising to the equivalent of 230 vessels of 30 000 cbm by 1981.

Obviously such high figures must not be taken too literally, but are an indication of the degree of uncertainty involved. What we can say is that in view of the worsening energy situation in the USA and growing demand in Japan it is reasonable to assume that a substantial demand for new LPG tankers will occur by the early 1980s, but probably not before then except for selected mid-range sizes.

To summarize, both LNG and LPG tankers are presently surplus and will generally remain so until the end of the decade. We expect a need for perhaps 20–30 new LPG tankers by the early 1980s, but not for new LNG tankers. By 1985 there should be a solid demand for new gas tankers of both types, barring some new energy source or economic disaster.

STEEL SHIPYARD CAPACITY

The next question concerns the capacity of existing steel shipyards to meet these gas tanker construction requirements. Our conclusion is that there is easily adequate steel shipyard capacity to meet the likely demand for gas tankers until the mid-1980s, and that efforts to market a concrete vessel will thus face very stiff competition until that time.

In our combined maximum cases we are theoretically projecting a requirement for new gas tankers reaching about 10 million dwt by 1985. By comparison, total tankers on order reached a peak in July 1974 of 195 220 635 dwt (1300 vessels).[2] By December 1976 this order book had declined, due mostly to cancellations, by nearly 150 000 000 dwt to 466 vessels totalling 45 293 383 dwt.[3] Assuming that these orders were spread over a five-year period we may, in a rather unsophisticated way, estimate that some 30 000 000 dwt of annual tanker delivery capacity is currently idle. We know that Japan alone is capable of turning out over 70 very large crude carriers per year,[4] equal to the new tonnage requirement of our maximum case.

While only a fraction of this yard capacity is normally dedicated to gas tankers, it is safe to assume that more than adequate steel yard space will be made available for any gas tanker demand by desperate yards until such time as orders for conventional tankers and other shipping bring yards again to near full employment.

When will that be? The present surplus of tankers of all types exceeds 100 000 000 dwt, with 351 ships totalling 31 011 000 dwt actually in lay-up.[5] It is generally agreed that, while conventional tankers of less than 60 000 dwt may show some new demand by 1980, larger tankers will be surplus at least until 1985. Furthermore, bulk and general carriers will be unable to absorb the surplus yard capacity. Bulk carriers are expected to be surplus until the early 1980s. General cargo vessels may be short by 1980, but the requirement for new vessels is expected to total only about 14 million dwt.[6]

Our conclusion is that capacity to build gas carriers in conventional yards will not be short until at least the mid-1980s.

COMPETITIVE PRECONDITIONS

Let us now move to the commercial or competitive preconditions which must be met before concrete ships will be successfully marketed, granted that this may be of only academic interest for several more years. We accept in advance that a safe, comfortable, and reliable concrete vessel can be designed. This, however, is only the beginning.

From a shipowner's point of view a concrete gas carrier will remain an unproven concept regardless of how many remarks are made about the concrete ships built during the World Wars, the wide use of concrete offshore structures, or the long life of the concrete hull. Shipowners are rather conservative with regard to basic technological innovation in ship construction, and with good reason in view of the costs and risks involved.

For anyone attempting to market a concrete gas carrier a number of elements must be met, in our opinion, to achieve success.

(*a*) A prototype must be built. It is going to be very difficult for an owner to undertake an investment which may be US$50–100 million without a prototype.
(*b*) The finished vessel cost must be substantially lower than that for a steel ship. It is doubtful that a differential of 10–15% will be adequate incentive.
(*c*) The construction time must be competitive.
(*d*) Firm price contracts with limited escalation will be necessary.
(*e*) The ship must be contracted on a turnkey basis with all the responsibility being placed on the yard. In other words, a contract to 'manage' a number of subcontractors will not be acceptable in most cases.

This last point poses a very interesting question in the development of concrete shipyards. If a company usually involved in civil engineering construction has an interest in this field it will have to become involved in two very new areas:

(*a*) naval architecture and propulsion;
(*b*) gas containment and handling.

One could find a contract for a concrete gas carrier where only 30–40% of the overall contract amount is civil engineering construction. It is easy to foresee the risk of such a venture.

CONCLUSION

In all countries that undertook construction of concrete vessels during the World Wars, two main elements were present: there was a shortage of steel, and the demand was not generated by market condition but for war reasons. Most of the ships were not competitive after the war ended. Some of them were built to carry such commodities as coal, and the heavy hulls rendered them uncompetitive. Specific gravities of liquefied gas range from 0·7 to 0·42, which compensates for the weight of the hull. But now there is no shortage of steel, and the demand must be generated in stiff competition with steel yards which need the work desperately.

If concrete gas ships are to be built for commercial purposes the following are the minimum requirements we foresee.

(*a*) The market for gas carriers must improve dramatically, resulting in substantial new demand for gas ships.

(*b*) A prototype must be built and operated to remove all questions about technical feasibility.

(*c*) Concrete shipyards should be created that meet the contractual needs of shipowners including a substantial cost saving v. steel construction.

In our opinion shipowners will not turn to concrete gas ships until market conditions force them to do so. When this happens the concrete industry will be ready to grasp the opportunity.

REFERENCES

1. Hay N. E. World LP-gas tanker supply and demand 1974–78. Paper presented at the 1974 meeting of the US National LP-Gas Association.
2. Ships on order end September 1976. *Motor Ship*, 1976, Oct., 8.
3. Ships on order end December 1976. *Motor Ship*, 1977, Jan., 8.
4. *International petroleum encyclopedia*. Petroleum Publishing Co., Tulsa, 1975, 424.
5. H. P. Drewry (Shipping Consultants) Ltd, London. Private communication, Feb. 1977.
6. Terminal Operators Ltd. The future for Greek shipping up to 1990. In Ships on order end September 1976. *Motor Ship*, 1976, Oct., 6.

BIBLIOGRAPHY

Gerwick B. C. *Prestressed concrete ocean structures and ships*. Prestressed Concrete Institute, Chicago, 1975.

Concrete afloat. Thomas Telford Ltd, London, 1977, 167–178

The dock is longitudinally divided into eight elements, which will be built separately, partially prestressed, and floated; the assembly of the elements will involve a further prestressing. In a model to represent the behaviour of a floating dock under ship loading, ship and dock are simulated by beams having the same overall stiffness, and rods representing the blocks are inserted between them; the lower beam rests on springs which simulate the response of the water. Finite element analysis methods were used to obtain the relationship between forces and deformations for the individual elements of the dock and the structure as a whole. Results obtained from the analytical model are presented. The analysis of creep and shrinkage effects is discussed.

Genoa's floating prestressed concrete dock: design

Prof. Ing. F. LEVI*

The construction of a new dock for supertankers was decided by the Consorzio Autonomo del Porto di Genova. The scarcity of space determined the choice of a floating dock, and the prestressed concrete solution was adopted as it emerged as the most economical among the many proposed in an international bid. Fig. 1 shows the main geometrical characteristics of the dock: overall length 342·82 m, overall width 78·25 m, internal width 62·91 m. The lifting capacity is of 100 000 tons, with a supplementary (reserve) lifting force of 15 000 tons.

A reticular spatial structure forms the horizontal caisson, while the two longitudinal walls are tubular concrete structures internally strengthened by transverse frames.

The sections of the dock are described in Fig. 2, while Fig. 3 shows a partial wooden model and a detail of a node.

Steel tubes in a tetrahedral arrangement form the internal structure of caissons. The upper nodes of the lattice are connected in the longitudinal sense by stiff steel girders, on which bear the blocks supporting the ship. The lower nodes of the lattice are connected by prestressed concrete beams. The prestressing tendons run evenly distributed inside the slabs in both orthogonal directions.

The dock is longitudinally divided into eight elements, or caissons, which may be independently ballasted. Each element will be built separately, partially prestressed, then floated and assembled to the others. A final longitudinal prestressing will give the structure its full continuity.

* Istituto di Scienza delle Costruzioni, Turin.

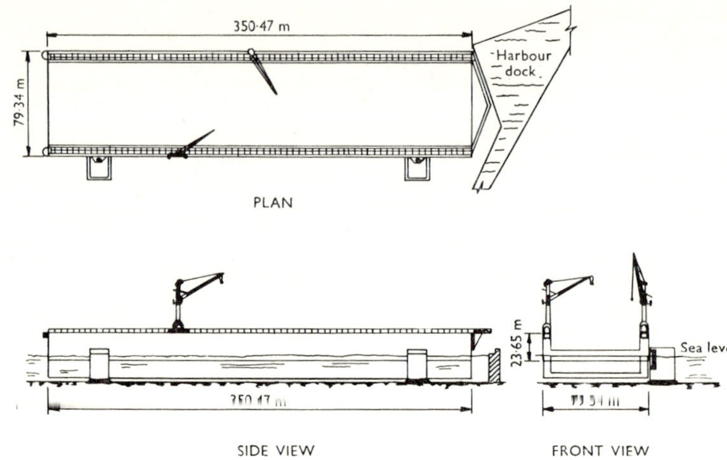
350·47 m
79·34 m
Harbour dock
PLAN
23·65 m
Sea level
SIDE VIEW
FRONT VIEW

Fig. 1

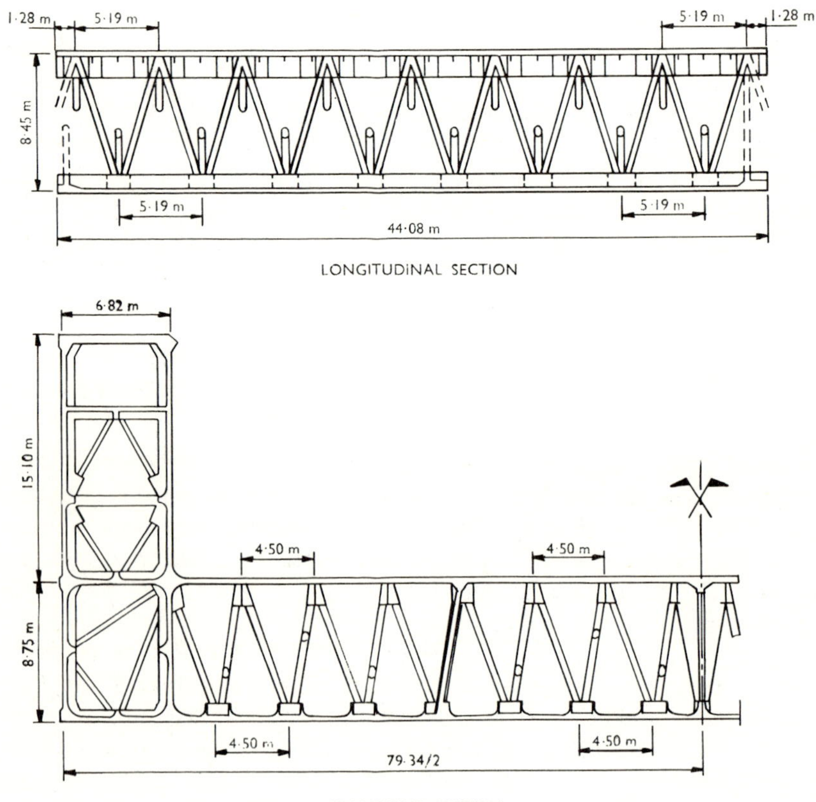
1·28 m
5·19 m
5·19 m
1·28 m
8·45 m
5·19 m
5·19 m
44·08 m
LONGITUDINAL SECTION
6·82 m
15·10 m
8·75 m
4·50 m
4·50 m
4·50 m
4·50 m
79·34/2
TRANSVERSE SECTION

Fig. 2

BEHAVIOUR OF FLOATING DOCKS UNDER SHIP LOADING

A floating dock carrying a ship is a rather complicated system composed of several elements coupled in parallel. The transmission of the ship's weight to the dock is strongly influenced by the ship's own local and global stiffness as well as by that of the dock. Furthermore, the way the ship's weight is distributed along the dock is affected by the presence of wooden and rubber blocks which bear up on the ship's hull. Lastly, the upward thrust of water on the bottom slab is not just related to the mean equilibrium level but varies from point to point depending on the dock's deformation.

A simple yet comprehensive model for dealing with the problem is illustrated in Fig. 4. Ship and dock are simulated in the longitudinal sense by beams having the same overall stiffness, and a series of rods, representing the blocks, is inserted between them. The lower beam rests on springs which simulate the response of the water.

The ship's weight on the dock may have a very variable distribution in the longitudinal sense, depending on the characteristics of the ship and on the load present. Several limit distributions were considered, in examining the most unfavourable stresses on the dock. Fig. 5 shows a typical result of these analyses. The diagram of the load transmitted by the blocks

Fig. 3

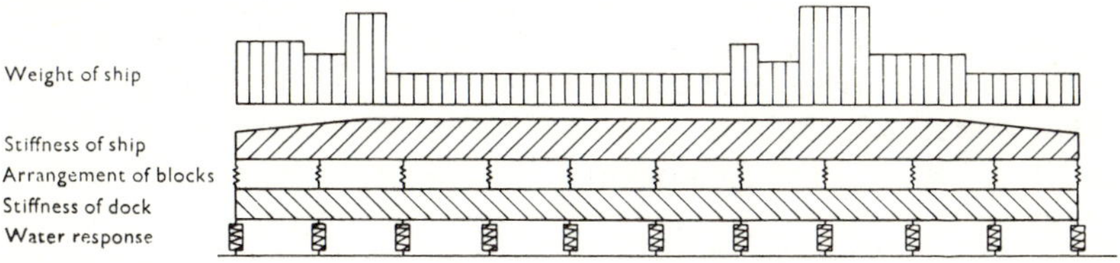

Fig. 4

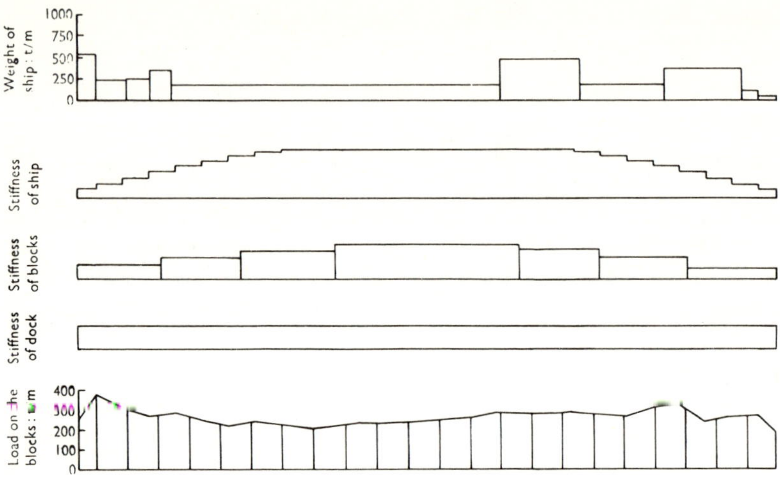

Fig. 5

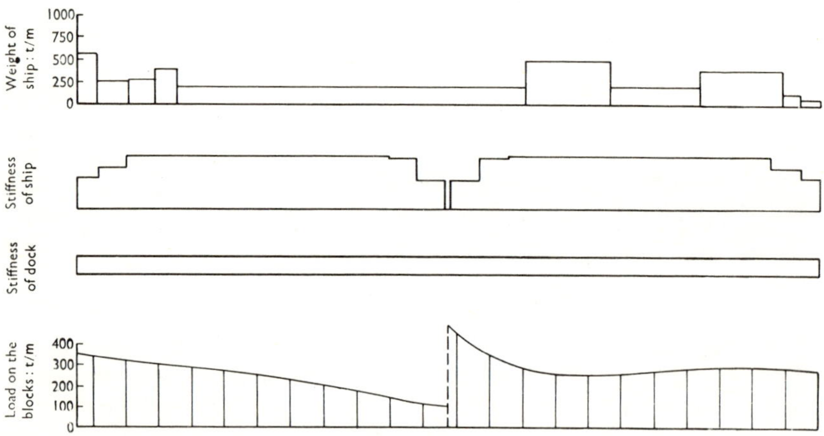

Fig. 6

appears much smoothed by the ship's stiffness as well as by the deformability of the blocks.

The case has also been examined of a ship cut in the middle for repair: the ship's weight and stiffness distributions are shown in Fig. 6, with the corresponding loading on the blocks: the unavoidable pressure concentration at the ends dictated special provisions for block arrangement.

Transverse deformability of the dock is not included in the interaction model of Fig. 4; its effect would be a further smoothing of the loading diagrams.

The structural model of the dock actually adopted for the analysis is capable of solving the interaction problem by taking into account

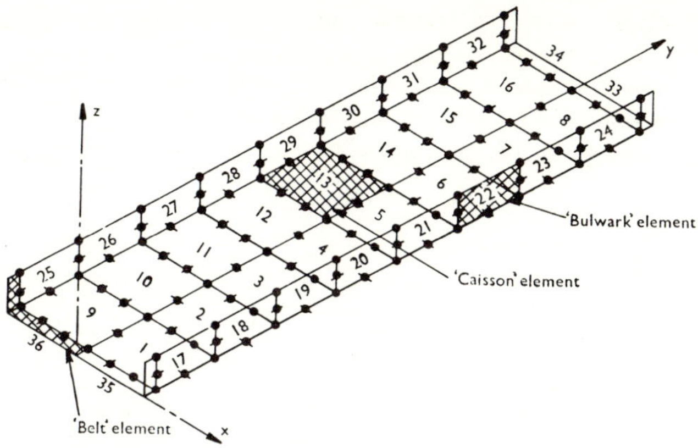

Fig. 7

transverse as well as longitudinal deformability. Uncertainty about other aspects would, however, render the accuracy of that kind of analysis somewhat illusory. The load–deformation relationship of the blocks has a high degree of dispersion, the contact with the ship's hull being by means of a wooden pad.

Furthermore, the ship's transverse bulkheads and frames introduce a strong variability of the local stiffness of the hull: the load exerted on the blocks tends to concentrate in the proximity of those elements.

Based on the considerations above, the more complex analysis was deemed unnecessary, and the results from the simplified model were adopted.

STRUCTURAL MODEL AND FINITE ELEMENT ANALYSIS

The particular structure of the dock makes unsuitable a continuum method of analysis. Some kind of discretization has thus to be envisaged, and the first step is the choice of a convenient type and number of parameters. Displacements along longitudinal and transverse axes were taken to describe the overall deformed shape; nodal parameters, located on the axes, are used to yield a continuous deformation line for them.

Figure 7 shows the axes actually adopted. Each of the resulting elements possesses several contour nodes.

The basic feature of the analysis is the separate study of the behaviour and the subsequent assemblage of the elements. Three types of element exist: 'caisson', 'bulwark' and 'belt'. For each type, which is itself a structure smaller than but as complex as the whole dock, the analysis consists in finding the relationship between force and deformation parameters along the contour. The 'shape functions' associated with each parameter are the Lagrangian polynomials often used for finite element formulations.

The analysis of the elements consists in finding the relationship between nodal force and displacement parameters using a finite element discretization. The schemes adopted for caisson and bulwark are shown in Fig. 8. A finer mesh has been applied for detailed analysis of flexural effects on limited areas of the slabs.

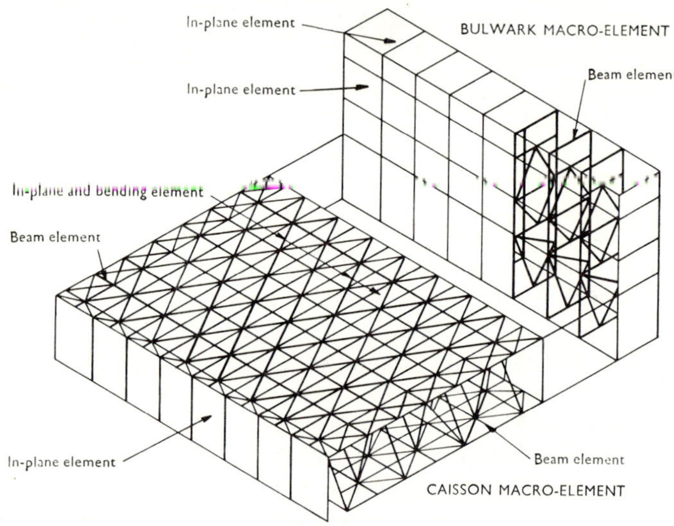

Fig. 8

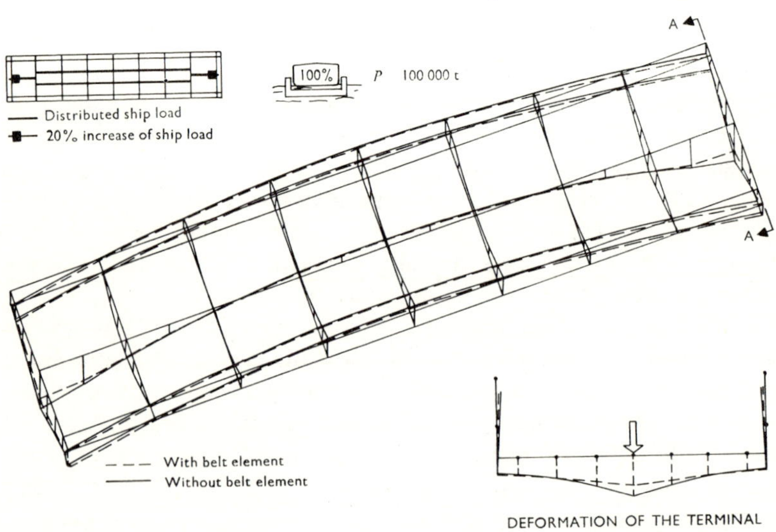

Fig. 9

BEHAVIOUR OF THE DOCK UNDER DESIGN SHIP LOADS

The analytical model described has been used both for design and for the final stage of analysis. By introducing successive modifications, the effect of structural changes could be examined. As the most significant example, the need for a belt element emerged from the analysis of some loading conditions. As seen in Fig. 9, the belt element has a stiffening effect in the transverse direction, notably reducing the deformation in the terminal zones. The stresses, which would be unacceptably high for a terminal caisson element, are easily taken by the additional element.

With the structure completely defined, the behaviour of the dock has been investigated for a variety of loadings, summarized in Fig. 10. The

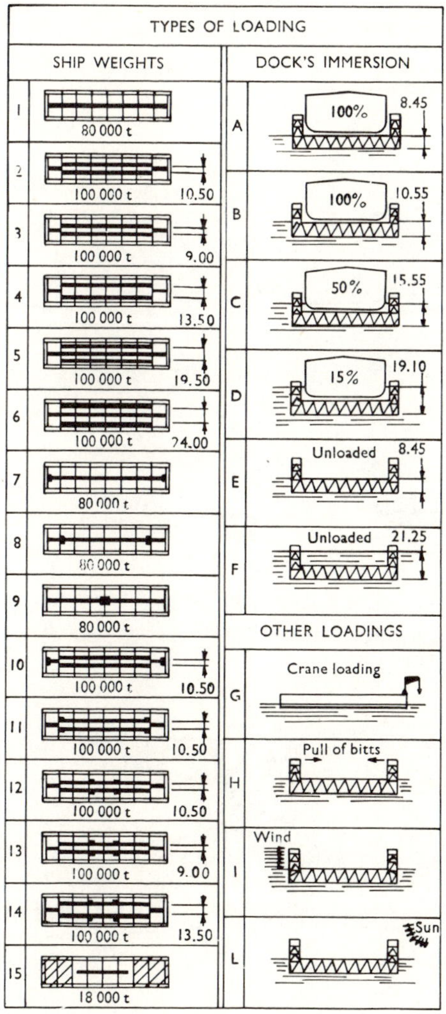

Fig. 10

most significant contract ship loadings, with their respective block arrangements, are shown in the left column of Fig. 10; on the right are the dock immersions and the other effects considered. Fifty-four combinations were analysed up to the level of generalized parameters; for 20 of these the stresses in any point have also been worked out.

An illustrative description of the essential results of the analysis is presented. Fig. 11 shows two deformed transverse sections of the dock, one near the belt, the other at middle length, for a typical ship loading. The comparison gives an idea of the three-dimensional behaviour of the structure as a function of the spatial distributions of both the ship's weight and the dock's stiffness.

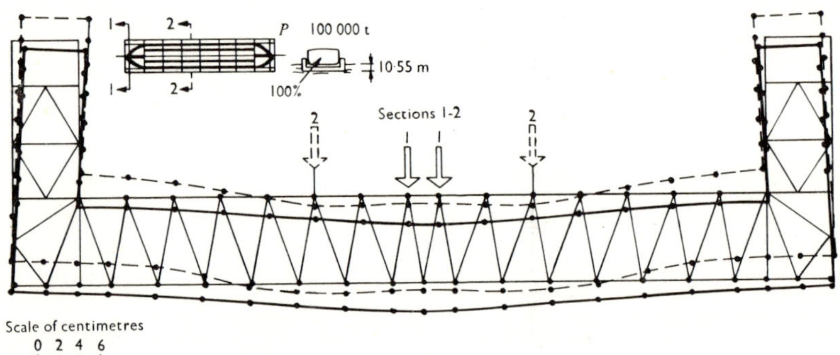

Fig. 11

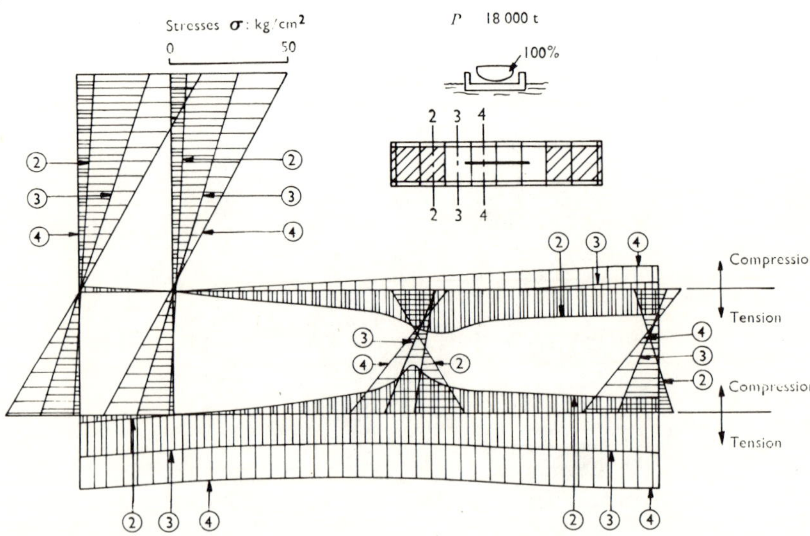

Fig. 12

The same aspect is also illustrated by the diagrams of longitudinal membrane stresses on caisson and bulwark elements. It can be seen from Fig. 12 that the middle section of the dock behaves essentially as a beam section, with a linear stress profile and a neutral axis nearly coinciding with the upper slab of the caisson. The stress states of bulwark and caisson become more and more independent at increasing distances from the middle section.

Flexural stresses are present on upper and lower caisson slabs, due to overall deformation and to locally distributed loads. For a detailed analysis of such effects, limited portions have been extracted and a very

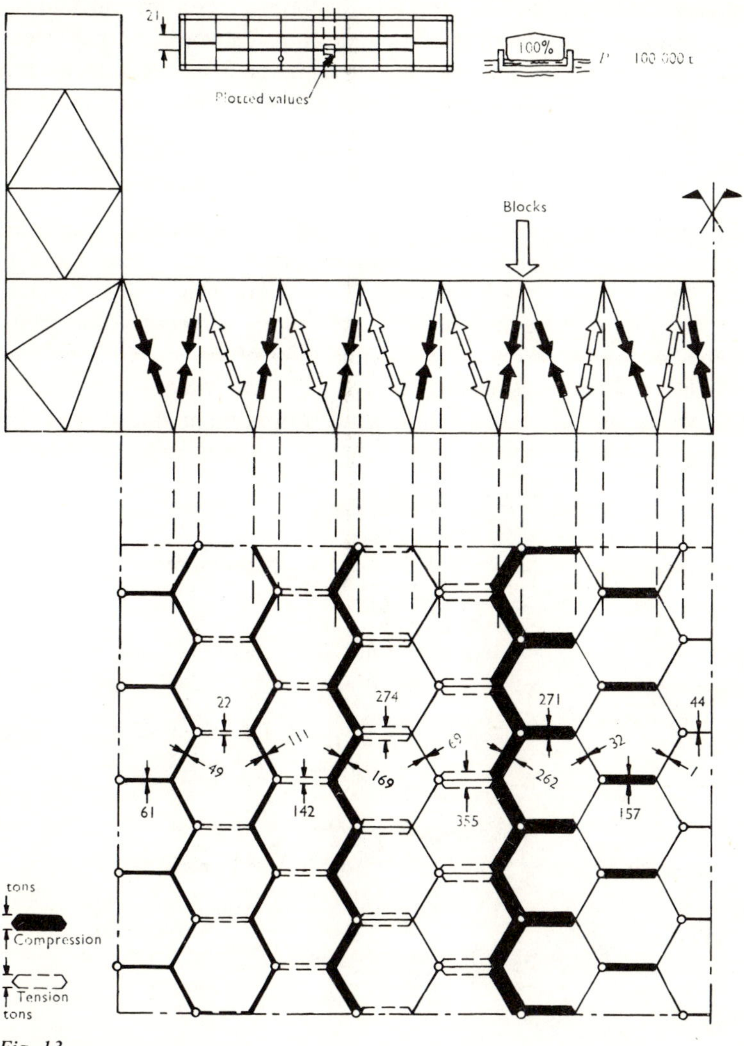

Fig. 13

fine mesh used for them. The nodal displacements obtained with the global mesh have been applied to the contour nodes of the new one.

The inclined struts of the caissons are other important structural elements: Fig. 13 shows the distribution of normal forces on them for a longitudinal segment of dock. Bending and torsional moments, deriving from the hypothesis of rigid nodes, were of secondary importance in the dimensioning of these elements.

The diagram of transverse membrane stresses on the upper and lower slab is presented in Fig. 14. The internal bending moment diagram is obtained by multiplying the stress resultants (product of stresses by the slab thicknesses) by the height of the caisson. Several tests for calculation errors were made, by checking the equilibrium between internal stress resultants and external actions, in the transverse as well as in the longitudinal sense: the accuracy of analysis was within the limits of 2–3%.

ANALYSIS OF CREEP AND SHRINKAGE EFFECTS

In order to meet the stress requirements at all points of the structure an accurate calibration of prestressing was necessary, taking into account the long term behaviour of concrete. This study was rather difficult because of the complexity of the structure, in which numerous heterogeneities were present, and the adoption of several construction phases. The problem was treated in two different ways. A computer program was developed, based on the same structural model adopted for

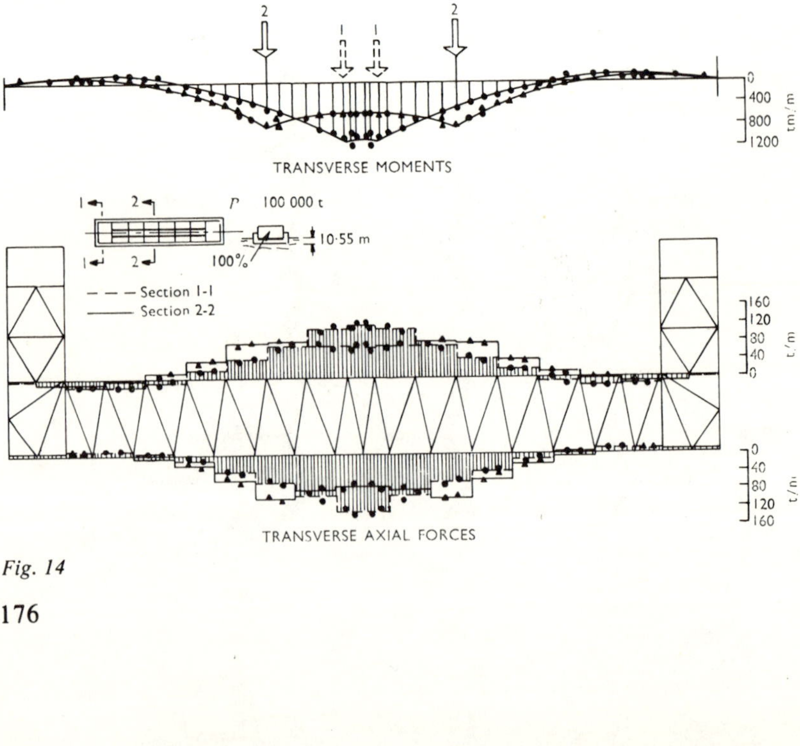

Fig. 14

the analysis of service load effects, but considering the stiffness matrix of the concrete structure alone to be able to describe the loss of prestressing and the migration of stresses from concrete to steel; this approach required the introduction of several approximations. Another control was made by a step-by-step procedure based on a different geometrical hypothesis covering the extreme conditions of collaboration between the bulwark and the floor.

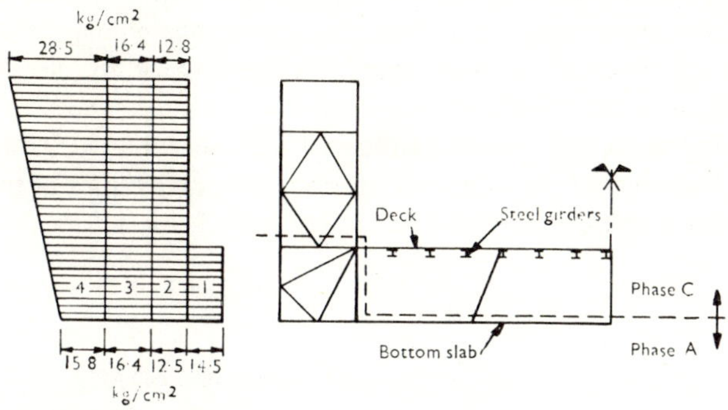

A: casting of bottom slab and vertical walls up to deck level
B: uniform prestressing (Pr. 1)
C: casting of the deck and the rest of the bulwark; the longitudinal girders are not bonded to deck slab
D: uniform prestressing of the entire structure (Pr. 2)
E: floating of the segment on basin and further prestressing (Pr. 3)
F: final assembly of segments by prestressing (Pr. 4) and bonding of girders to deck slab

Fig. 15

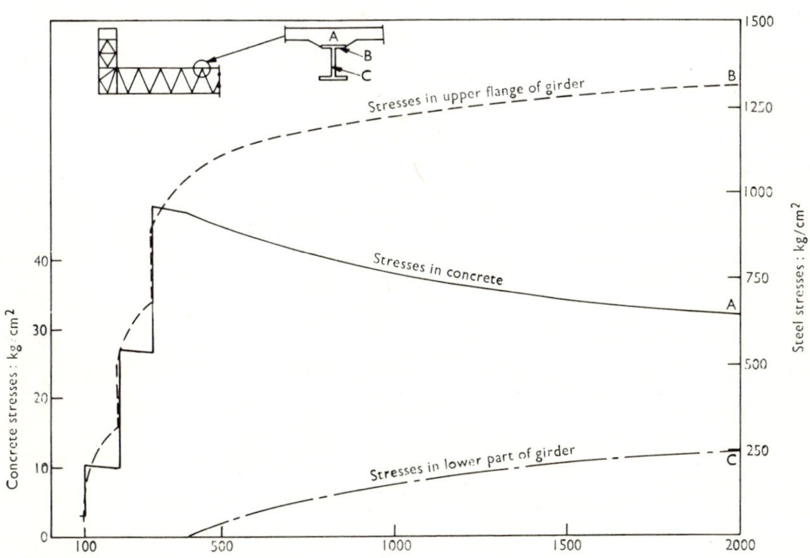

Fig. 16

The sequence of construction and prestressing phases for a segment of the dock are summarized in Fig. 15, in which the stresses induced in concrete at each phase are also indicated. Greater asymptotic values of creep and shrinkage deformations have been taken for the upper part of the structure, since the temperature and humidity conditions there are more variable.

In Fig. 16 the migration of the stresses from the concrete to the steel girders is represented. The vertical steps of the concrete stresses are the effects of the prestressing sequence. An increase is evident in the rate of reduction of concrete stresses after the welding of the girders to the flanges already bonded to concrete.

The delay in welding, which in fact will be larger than 400 days as supposed in the calculation, is beneficial since the final values of the stresses on girders are much reduced.

Concrete afloat. Thomas Telford Ltd, London, 1977, 179–194

The Consorzio Autonomo del Porto di Genova (the Genoa Port Authority) has entrusted SOGENE with the construction of a floating dock of 350 000 dwt. The principal dimensions are as follows: length 350·17 m, width 79·04 m and total height 23·85 m. The floating dock, with a U shaped middle section, consists essentially of a floor formed by two lightweight prestressed concrete slabs straightened by a tubular lattice and by two bulwarks, also in lightweight prestressed concrete, formed by a framed box structure. Thus a weight reduction with regard to the normal cellular structures is obtained, realizing a scheme particularly suitable for the construction of floating works such as pontoons, platforms etc. Due to the peculiarity of the work, the study of the statical behaviour, carried out with finite element methods, has been controlled and integrated with tests on models. The floating dock is presently under construction. In a special dry dock, eight identical elements are built separately, then launched and progressively joined together to obtain a monolithic final structure.

Genoa's floating prestressed concrete dock: construction

Dott. Ing. M. COSTANTINO*

The prestressed concrete floating dock, at present under construction in Genoa, is of such a size that it cannot be constructed in one piece.

The structure of the floating dock consists of concrete slabs connected by a space framework, continuous both in the transverse and in the longitudinal direction. This structural continuity must be maintained throughout construction to avoid points of discontinuity in the structure where considerable concentrations of tension stresses could occur.

The number of constituent elements of which the floating dock will consist depends on the size of the dry dock in which construction of the elements will be carried out. The dry dock in Genoa (Fig. 1) is surrounded by concrete caissons, and has inside dimensions of 50 m × 90 m, thus allowing for the construction of the six 43 m long and two 46 m long dock elements, each being 79·04 m wide.

The dry dock can be closed at one and by a metal ship gate; during flooding it allows the entry of water over all the contact surfaces between the floor of the dry dock and the base of the floating dock element under construction, thus avoiding an abrupt separation of the latter during launching.

The height of the dry dock is such as to allow for a depth of 1·3 m water between the lower surface of the completed floating precast element and the top surface of the base of the dry dock. Such a clearance is required to allow, during floating, the introduction of a number of steel

* Director, Società Generale per Lavori e Pubbliche Utilità SpA.

support blocks placed at the stiffest sections of the dock element, so that before final launching the bottom and parts of the walls of the floating dock element which will remain permanently under water can be subjected to inspection (Fig. 2).

Fig. 1. The dry dock and a completed element of the floating dock

Fig. 2. Steel blocks placed under the bottom slab of a floating dock element to facilitate inspection

The design of arrangement of the ballast tanks of the floating dock is such that the ends of each of the eight dock elements are formed by wall structures which can be either temporary or permanent. Though both types of end wall are made of steel, they are of differing construction. The permanent end walls (Fig. 3) consist of steel plates which will not be affected by any transverse deformations of the dock under load as the plates are surrounded by neoprene joints; owing to the stiffness of the plates the end walls do not interfere with the stress distribution in the structure. The temporary walls are built in sections which, once two elements of the dock have been joined, can be extracted and used later in the construction.

Longitudinal continuity of the floor structure is attained by means of the space framework by welding steel rods across two contiguous elements and welding steel beams connecting the upper junction units. Finally, after the concrete is placed, stressing of the cables creates structural continuity, assuring substantially identical behaviour of this section with the others. Owing to the monolithic nature of the structure, the result of prestressing and its structural continuity, the floating dock can be safely towed in open seas with 5 m high waves 80 m long.

The time required for building the floating dock depends on the number of elements it comprises and therefore on the size of the dry dock in which the elements are being built. For the Genoa dock (Fig. 4) the time envisaged for the completion of such structures allows for the construction of eight elements with an average time of six months for each

Fig. 3. Permanent water-excluding end wall

element. The number of man-hours required for the completion of each element is 340 000. The dockyard is so organized that production can be industrialized to the utmost. The lightweight aggregate is stored in a place well protected from humidity. Two batching plants ensure continuity of casting which is effected by four cranes or by a conveyor belt (Fig. 5). The preparation of the prestressing cables and the pre-assembly of the ordinary reinforcement are carried out in covered areas.

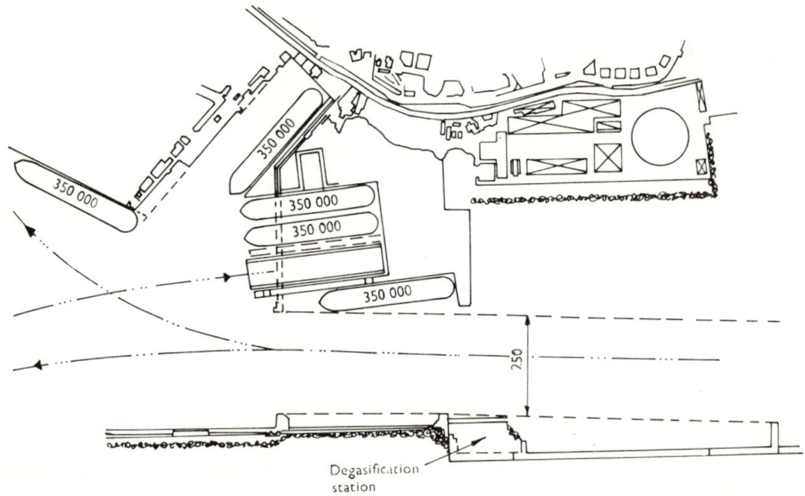

Fig. 4. Arrangement of the floating dock area

Fig. 5. Belt conveyor for casting concrete

The construction of the structural steelwork components and the assembly of the joints are carried out in a workshop on the site which is equipped among other things with a sheet steel cutter and a press. The formwork for the sides has been so designed that it can be used successively by making it roll along like slip forms. During the construction of the sides, the vertical prestressing cables are stored rolled up in drums (Fig. 6).

MATERIALS

The design of the structure required that the concrete should be impermeable and that it should have a density of 1950 kg/m^3 and $f_{ck} \geqslant 45$ MPa. Preliminary tests showed that such properties could be obtained with a concrete mix made up, for each cubic metre of finished concrete, of 400 kg special 525 blast-furnance Portland cement, 490 kg of $\leqslant 4$ mm size sand and 860 kg of ARES 9 structural lightweight aggregate, the apparent specific weight of which is about 1600 kg/m^3 for a grading of 30–50 mm (Fig. 7).

A plasticizer was added to allow for a water/cement ratio of about 0·40 during casting. In the concrete so obtained, the texture was uniform, the aggregates were evenly distributed and voids were insignificant (Fig. 8). Laboratory tests with pressures of 20 atm showed that the resistance of the concrete against the penetration of both water and chlorine ions was satisfactory. Compression tests on samples during construction (Fig. 9) gave a crushing strength of $f_{ck} > 47{\cdot}5$ MPa; the tensile strength obtained with the splitting test exceeded 2·5 MPa; creep and shrinkage tests

Fig. 6. Slip forms and cable drums used for construction of the vertical bulwark

gave similar values to those usually obtained with ordinary concrete of the same quality; and the tangent and secant elastic moduli reached 21 000 MPa and 17 000 MPa respectively. The successful use of this particular concrete mix depends on accurate knowledge of its setting and hardening properties related to the quantity of water in the mix, including the quantity of water contained in the aggregate (i.e., the effective water/cement ratio).

The prestressing cables consist of stabilized steel wires with a breaking stress of not less than 1650 MPa. The system used in Genoa was that of

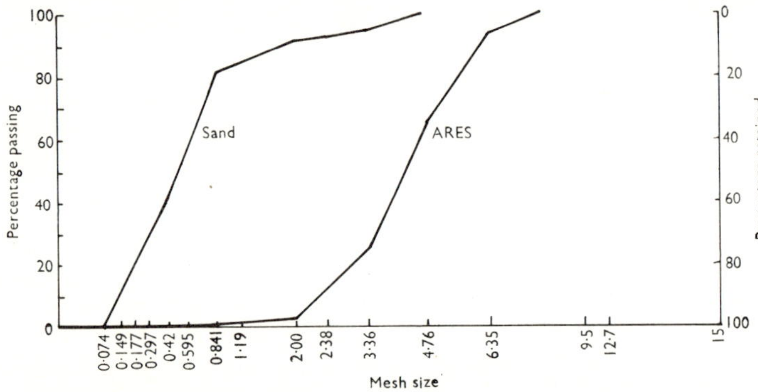

Fig. 7. Grading curves for sand and ARES 9 lightweight aggregate

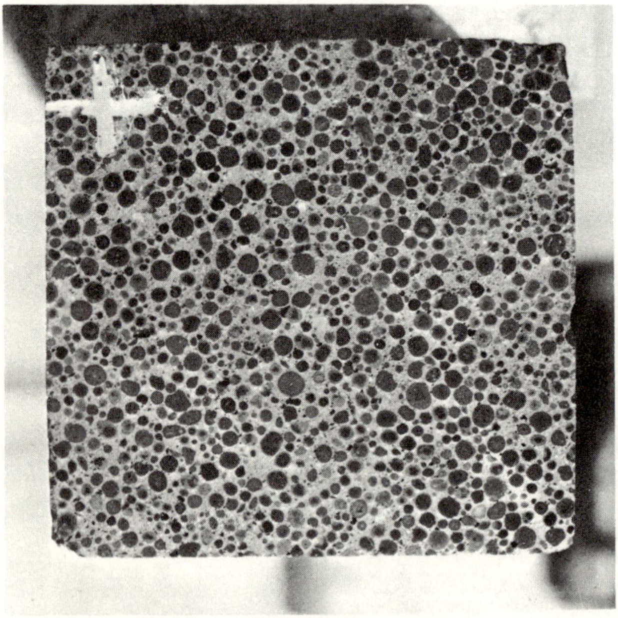

Fig. 8. Cut face of a lightweight concrete cube

BBRV with sheet metal sheathing and with steel pipes to avoid the grouting of the first phase affecting the cables of the second and third phases. The subsidiary reinforcement is composed of high bond bars having a yield point higher than 440 MPa and electrically welded square mesh reinforcement.

Junction units form the connection between diagonals of the

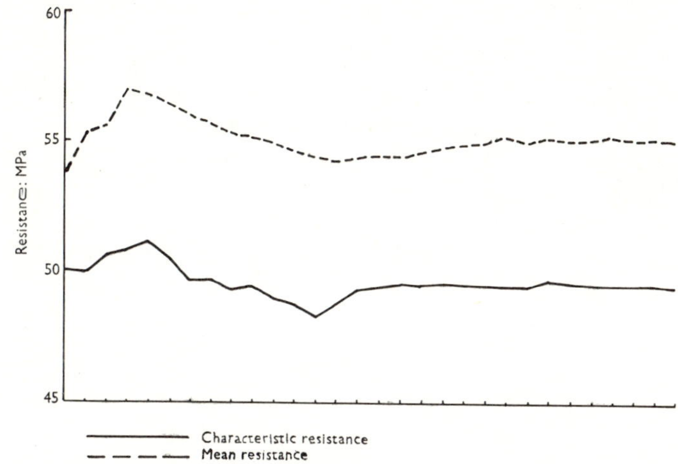

Fig. 9. Characteristic and mean strengths of the concrete at 28 days

Fig. 10. Lower steel junction units

Fig. 11. Upper steel junction units for the longitudinal beams

Fig. 12. Tubular steel structure

longitudinal beams and the upper and lower slabs of the floating dock. The lower junctions (Fig. 10) each connect three diagonals to the lower slab and are made of one single cast steel block weighing 550 kg. Special design studies and tests led to the adoption of the present shape and size of the junction units as well as the choice of riveting for their connection with the prestressed slab. The upper junction units are more complicated, because they also include the longitudinal beam (Fig. 11): at first they were envisaged as cast steel components but because of the difficulties encountered in obtaining the block thicknesses required to resist high stresses, it was decided to use a welded, fabricated steelwork construction.

The truss members which form the space frame structure (Fig. 12) are made of seamless smooth steel pipes with a constant outside diameter of 406 mm. The thickness of their walls ranges from 10 mm to 28 mm according to the stresses to be resisted. The mechanical properties of the steel are $f_s = 550$ MPa, $S \geqslant 3{\cdot}6$ MPa and $A \geqslant 21\%$.

The upper beams perform the task of transferring loads acting on the top deck of the floating dock, namely the reactions of the support blocks, dead weight, water load etc., to the upper junctions of the space frame. Internal forces also arise as a result of prestressing, by the creep and shrinkage of the concrete slab acting compositely with the upper flanges of the beams by means of steel shear connector pins welded electrically to the flanges.

To contain these effects it was arranged to ensure the continuity of the webs and lower flange of the beams in a phase following the prestressing of the slabs. In this way only the upper flat arches are integrally involved in the prestressing and the slow deformation of the slab.

MODEL TESTS

Before beginning the work a series of tests was carried out on models, to check the results of calculations or to obtain information which could not be derived from calculations.

A first model reproduced a 12 m long section of the dock for its whole width on a scale of 1:5 (Fig. 13). The model was made with the same materials as those which were to be used on the job. In order to simulate the actual hydrostatic pressure, the model was placed on rubber bags filled with water under pressure, while the equivalent load representing the weight of the ship was imposed with jacks acting on the top deck against ballast. The tests gave results which were in agreement with the calculations for the overall stress conditions. The local effects due to the hydrostatic thrust acting on the base slab, however, were not reproduced accurately because of the shear stress induced by the bags when dilated by hydrostatic pressure and because of the edge effects due to the limited length of the model. The structure responded elastically up to loads one

Fig. 13. Testing equipment for 1:5 scale model

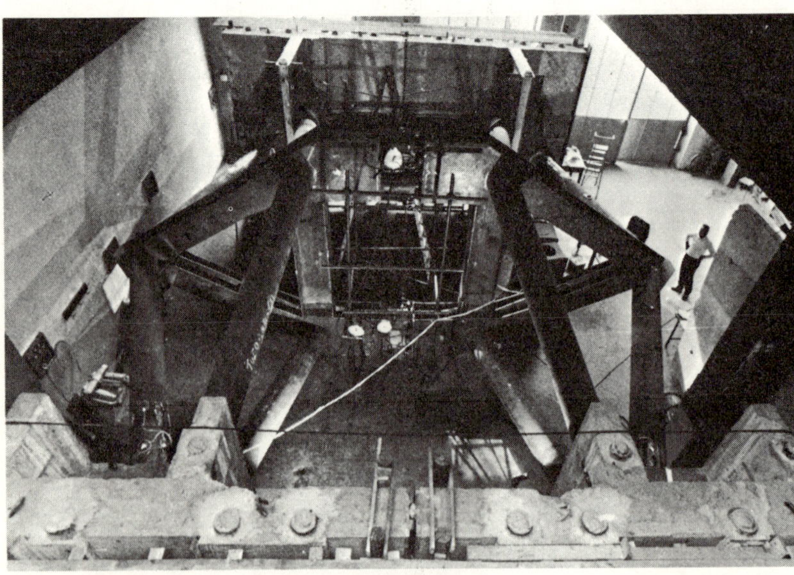

Fig. 14. Testing full-size model of junction units

third higher than the working design loads (the maximum obtainable with the available testing equipment) and residual deformations were within the limits of experimental error. By modifying the loading and support conditions (i.e., by resting the ends of the model on fixed supports) the loading test could be carried to destruction. During that test the behaviour of the model showed excellent conformity with theory, and even after cracking occurred in the lower slab the model continued to behave in a completely satisfactory way, showing the existence of ample safety margins and an adequate capacity of recovery.

Tests were also carried out on two additional models to verify the behaviour of the junction units (Fig. 14). The upper junctions in the models differed: one consisted of a single cast steel body while the other was of fabricated welded structural steelwork construction. Tests on full-size models allowed an assessment of the behaviour of the junction units with sufficient approximation to the severest conditions of service by imposing forces half-way along the length of the diagonal struts with hydraulic jacks and allowing the junctions to rotate by means of a hinged joint in the slabs. The following tests were made on the two models: magnetoscopic examination and ultrasonic testing of the welds before and after the loading tests; static loading tests using fragile-lacquer tensio-paints; static loading tests taking deformations and displacements up to a load 15% higher than the maximum design load; fatigue test under pulsating loads; and thrust load test at the maximum value obtainable. The tests showed that the strains measured during the experiments with different instruments were similar; the magnitude of the calculated and experimentally obtained tensile stresses was similar; and the behaviour of the structures under oligocyclic fatigue loading was satisfactory, and perfect elasticity and linearity of behaviour were retained after the test.

Other full-size models were also tested: tests were made to study the effects of the deformability of the steel shear connector pins on the state of tension in the slab and on the magnitude of the shear stresses; and tests were made on the anchorage units at the cable ends to study the distribution of the prestress and the effectiveness of the enveloping transverse reinforcement.

The space frame structure used for the dock has been used for floating loading pontoons. Junctions of a very simplified type were tested on full-size models. Besides checking conformity with the results of the calculations, the resistance to intermittent stresses of the structures in the sea was verified successfully through fatigue tests representing cycles of several hundreds of miles on the model.

REPAIR TECHNIQUES

In preparing the designs for such an important project it was imperative to provide for the possibility of repairing the structure should this need arise at any time.

Such repairs could become necessary as a result of two distinct contingencies. The first is local faults in workmanship: it seemed probable that in casting such vast areas of concrete slab, 300 mm thick, some im-

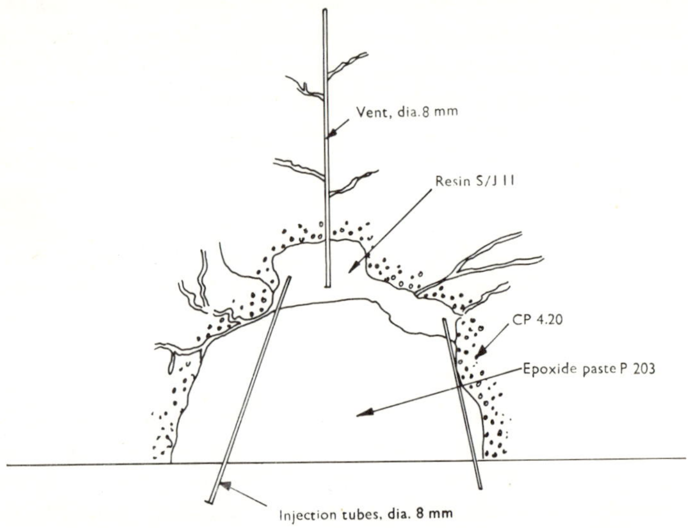

Fig. 15. Resin injection technique

Fig. 16. Slab repair tests

perfections might occur which could decrease the resistance of the structure to water penetration. The second contingency requiring repairs is that of unforseeable accidents, such as the collision of a ship with the sides of the dock.

In the first case, the repair technique envisaged is the use of cement grout or resin injections; these methods are well known and have been used with success on numerous occasions. This technique has been used on the present floating dock project for filling in of several zones of porous concrete.

Owing to the failure of the aggregate dosing system of the batching plant which occurred on the first day of casting, the bottom slab of the very first element built contained about 100 m^3 of concrete which did not conform with the mix design, in that the sand content varied and was generally lower than that which was required. This caused the formation of zones of porous concrete interposed with zones of well compacted concrete. Since these zones of porous material were small in relation to the total quantity of concrete cast, and as the cores taken from them showed that the concrete, although porous, had strength characteristics which were higher than those required in the design, epoxy resin injections were carried out over all the area concerned. The element affected was placed in the dry dock and the bottom was carefully inspected. A resin with very low viscosity, a long pot life and low heat development was then injected in the bottom slab from below (Fig. 15). These injections succeeded in completely blocking the pores in the concrete as shown by the concrete cores taken and by the local load and permeability tests made after the injections had been completed.

In the case of an accident, should a slab undergo damage and some cables break it is possible to replace the damaged concrete and provide additional stress by attaching special heads and lengths of cable to the original structure and by using an achorage with which it is possible to put a cable under tension from an intermediate point. Tests (Fig. 16) were made for this operation on full-size models by constructing slabs with a void in the centre and with broken cables. The broken cables were joined and then placed in tension after the central hole had been filled with concrete. After the repairs were made, deformation and loading tests on the slabs showed that it is possible to restore the prestress in the concrete effectively and therefore to carry out repairs which would bring the structures up to its original strength and stiffness.

ASSEMBLY OF THE ELEMENTS

Obtaining a correct combination of the eight elements built in the dry dock requires the following actions.

(*a*) Modest dimensional tolerances (thickness, positioning of the prestressing cables) must be respected.

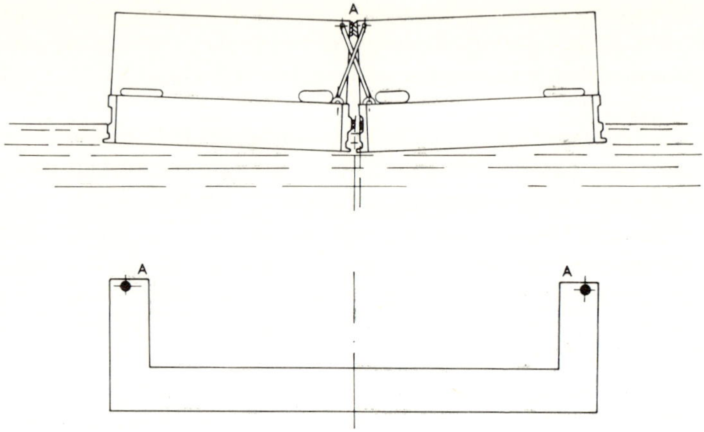

Fig. 17. Assembly operation (c)

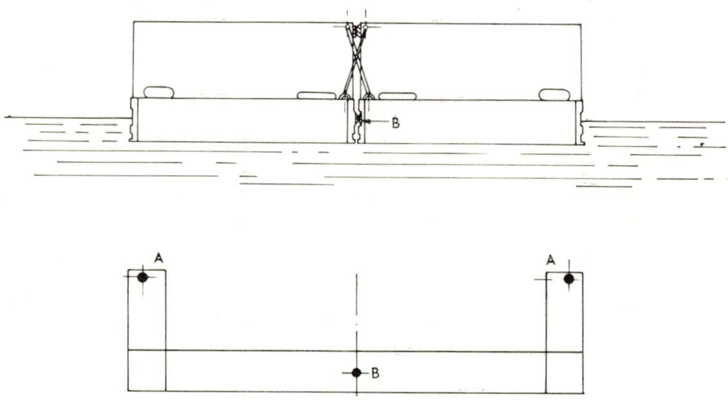

Fig. 18. Assembly operation (d)

Fig. 19. Three completed floating elements before assembly

(*b*) A sequence of operations must be followed which will reduce the risk of damage when bringing the elements into position, as damage could prejudice the completion of the successive phases of the combination of the elements.
(*c*) The elements must be positioned with great precision. This is made difficult not only by a different floating trim, but also by the presence of inevitable wave movements in the water.
(*d*) Even with these wave movements present, every other relative movement of the elements during the setting period (about three days) of the sealing concrete must be prevented.

The preparatory programme can be divided into the following phases:

(*a*) levelling, with suitable ballast, the trim of the elements to be joined in the water;
(*b*) bringing up the elements, first with a tug or by winching from land and then gradually with horizontal and vertical pulley-blocks up to a distance of 1 m from each other;
(*c*) suitable ballasting to produce a longitudinal inclination of the elements; the elements are then pulled together with pulley-blocks (Fig. 17) until contact occurs at points AA of the upper slab of the bulwark where steel plates have been placed;
(*d*) rotating the elements (Fig. 18) around the axis marked by points AA until contact has been achieved at point B of the central wall through suitable steel plates;
(*e*) final fixing of the elements with prestressing cables and wedges of special shapes placed at the central part, the upper slab of the bulwark and the internal walls of the bulwark;
(*f*) emptying the water from the space between the two elements after having placed a strip of rubber around the outside edges;
(*g*) casting non-shrinking concrete in the connection zones only;
(*h*) tensioning the prestressing cables in the connecting zones only;
(*i*) joining the steelwork components;
(*j*) filling with concrete all the remaining contact zones beween the elements and fully tensioning all cables, thus producing a completely monolithic structure.

PROGRAM FOR DOCKING PROCEDURE

A computer program has been evolved to assist with the correct utilization of the dock. When the distribution of the dead weight of the ship and of the loads carried are known, the program will indicate whether it is possible to dock the ship without exceeding the strength characteristics of the master section. If so, it will supply the distribution and quantity of ballast water needed so that the design loads on the beams and docking support blocks are not exceeded, the positioning of the docking blocks, and the trim and deformation of the dock. The results of the calculation are supplied diagrammatically by plotter as a weight diagram, a ship rigidity diagram, a ballast water diagram, a moment diagram, a deflections diagram, the load on the docking blocks, and the arrangement and spacing of the docking blocks. Automatic controls of the dock with a hydraulic, laser and optical system will give the on-board computer confirmation of conformity of actual deformations with those obtained from the program, and therefore the go-ahead for docking the ship.

CONCLUSION

Figure 19 shows three of the floating elements before assembly.

Concrete afloat. Thomas Telford Ltd, London, 1977, 195–204

Discussion

DEVELOPMENT OF THE CONCRETE HULL (MR MORGAN'S PAPER)

Mr Swift suggested that concrete might be a suitable hull material for pilot cutters, and showed some details of a design which was under consideration by the Mersey Docks and Harbour Board. He noted the high reluctance of the elder brethren of Trinity House to accept the use of floating concrete. Mr Morgan confirmed that this was also his own experience, and instanced the reluctance of Trinity House to use concrete navigational buoys.

Mr Impey drew attention to the possibilities of sprayed concrete and fibre-reinforced concrete. He showed a design for a pontoon using foamed polystyrene, with steel reinforcement, sandwiched between two layers of fibre-reinforced concrete, and stated that the estimated cost was £70 per cubic metre of vessel. In reply, Mr Morgan pointed out that processes involving a high labour content were often acceptable in developing countries but were less welcome in industrialized societies. Mass-production techniques could be used to reduce the costs of sprayed concrete or prestressed ferrocement, but only if the mass market existed.

Mr Morgan was then asked whether there were not significant differences between fibre-reinforced concrete and ferrocement; had fibre-reinforced concrete been proved satisfactory for marine applications, and was it not difficult to make? Mr Morgan thought this aspect, which applied also to polymer techniques, required further investigation. The techniques had been proved for land-based applications, but more research and development would be needed before they could be recommended without reservations for marine use.

PRESTRESSED CONCRETE PLATFORMS FOR PROCESS PLANTS (MR DERRINGTON'S PAPER)

Mr Impey queried the need for a cover of 40 mm. Mr Derrington replied that this was not an absolute value; the amount needed should be determined for each particular case.

Mr Bogdanović referred to the growth of algae on concrete. It had been found that this growth produces a surface film which gives added protection against corrosion. Danish tests had shown that corrosion was reduced by 50% when the algae were present. He thought that, provided the concrete was well compacted and free from cracking, the actual thickness of the cover was irrelevant.

Mr Pope asked whether the platforms were floated to their locations and then grounded, and why permanently floating platforms should not be used. Mr Derrington said that this depended on the case considered; in general it had been found that grounded structures were slightly more economical.

Dr Gjorv reverted to the question of cover and crack width. He noted that the algae–concrete–sea water reaction varied according to the quality of the concrete; fast disintegration had been found to occur in porous concrete but in dense concrete a thin layer of heavy salts was deposited which had been shown to be highly beneficial. The density and compaction of the concrete was of fundamental importance. There was no point in increasing the cover beyond about 60 mm, since above this value a constant diffusion of oxygen into the concrete occurred.

The size and distribution of cracks was highly important, since these lead to a different corrosion mechanism in which the embedded steel acts as a cathode for oxygen when exposed in the crack. A greater thickness of cover permits a bigger build-up of salts and corrosion products within the crack and is thus beneficial. A separate wire mesh, or fibres, should be provided to limit crack widths.

THE ROLE OF CLASSIFICATION (MR ALGATE'S PAPER)

Mr Morgan asked how the classification society was placed if one of its classified ships should break in two. Mr Algate answered that the investigation of accidents was carried out by the Department of Trade, who determined where the blame lay.

Mr Impey asked whether the charges of the classification society were paid by the shipowner. Mr Algate confirmed that this was so; the scale of survey fees was based on the work content.

Professor Gallin made a plea for a detached investigation of relative merits of steel and concrete hulls. He noted that shipowners were in a high-risk business, and disliked novelty; a strong incentive (at least a 10% increase in profit) was needed to persuade them to adopt a change. He hoped that non-committed data would be produced during the symposium.

Mr Algate was asked what was the attitude of the classifying authorities when new techniques were presented to them. He replied that they did their best to avoid causing unnecessary obstacles; their philosophy was to adapt to change.

CLASSIFICATION OF CONCRETE SHIPS (MR CAMPBELL'S PAPER)

Mr Derrington noted that the classification society rules were still in an embryonic stage. He drew attention to the different partial safety factors for the temporary and permanent states of a structure adopted in European codes. He asked what was the attitude of the ABS to the limit-state approach. Mr Campbell said that the ABS had reduced the factor of safety for the delivery period of the LPG refrigeration, storage and transfer terminal to the Java Sea. He hoped that people in the concrete industry would make their opinions known to the classification societies, whose experience was with steel structures; they depended on the industry for advice on concrete structures.

Mr Campbell was asked what the relationship was between Lloyds, ABS and other classification societies, and wheter a client had a free choice of classifying authority for a specific project. Mr Campbell said that the societies were in competition and that a client was free to go to whichever authority he chose.

Mr Worth commented on the difficulty of repairing damaged reinforced and prestressed concrete structures. Mr Campbell again emphasized that ABS had little experience with concrete structures and that they would be governed by whatever was laid down in the codes. He noted that ABS have seen split tendons repaired. Mr Carvill referred to concrete ships which were built during World War II and he emphasized the impressive condition which some of these structures still retained. He noted that one of these vessels had recently been up for sale at a price of more than $200 000.

OPERATIONAL EXPERIENCE WITH PRESTRESSED CONCRETE BARGES (MESSRS SARE AND YEE'S PAPER)

When questioned about the durability of prestressed concrete barges, Mr Yee replied that the water/cement ratio and the density of the concrete were the most important factors influencing corrosion. He added that cement with a C_3A content of 12% had been used. The ½ in. dia. prestressing wire used was designed to have a cover of 1½ in., although it was observed to be as low as ½ in. in places. Mr Bogdanović said he had found sulphate-resisting cements satisfactory for reducing corrosion.

Professor Gallin noted the economy of capital cost and of maintenance; he asked if the hatch covers and the rest of the steel superstructure lead to durability problems. Mr Yee said that the steel superstructure caused the same problems as steel hulls.

Mr Yee was asked by Mr Hetherington how he felt about patching damaged barges, and the effect on the strands. The author replied that he had never seen strands break under impact load; the only effect was cracking of the concrete. However, with the one barge which suffered a mine explosion, a 12 ft dia. hole resulted in the loss of 8–10 strands. Nevertheless, out of a total of 400, this number represented only a small percentage, and a steel plate was added to replace the lost tendons.

Dr Gjorv asked if marine growth created a problem after a long period of exposure. Mr Yee pointed out that the rivers where the barges operated had such a high pollution level that nothing could grow on the hulls.

LNG CONCRETE SHIP ECONOMICS (MR STANFORD'S PAPER)

Mr Ffooks disputed the maintenance costs and maintenance time which Mr Stanford specified as being necessary for steel ships. In his own experience with steel LNG carriers, they had tried to reduce maintenance levels by coating the tanks with coal tar. The ships were more than twelve years old and no steel repair had been necessary because of breakdown of paint protection. Mr Ffooks also said he found the single-hull concept novel and interesting. He wondered what might happen to the insulation system in the case of a substantial collision, and he asked to what extent concrete could cool without suffering too much damage.

Mr Stanford said that the company Mr Ffooks had been dealing with had a good maintenance record and hence their ships were in good condition. He did not think that the same standards were achieved elsewhere. He also felt that the insulation problems were just as great with steel ships having two shells.

Mr Bogdanović referred to work done in Barcelona on cryogenic tests on concrete to $-160°C$. The concrete strength of 40 N/mm^2 at ambient temperatures went up to 120 N/mm^2 on cooling. In addition it was found that the modulus of elasticity went up by 25%. He also said that he had tried non-corrosive steel linings for the tanks and he found them to be unnecessary. He said that 8% of water in the concrete turned to ice which gave a protective film. He questioned whether a lining was required at all.

Dr K. Finsterwalder described a technique developed for coupling damaged tendons.

Mr Manning reported a study on temperature after a breach had occurred in a cryogenic vessel. It was found that heat input from the surroundings raised the temperature to $-5°C$ at the breach.

GENERAL

Mr Furnes said that Det Norske Veritas had done work on LNG and concrete about two years ago. The DNV proposals for concrete ships had been presented elsewhere.[1]

Dr Drakel reported a series of tests carried out on four concrete cylinders at the Stevin Laboratory of the Delft University of Technology.

The experiments were designed to simulate the collision of a ship against a vertical concrete tower of an offshore gravity structure. The time taken to reach the failure load in some tests was less than 2 s; the corresponding time period on other tests was of the order of 1 h. All the cylinders failed in punching shear. The cylinders were 3 m long, and had an outside diameter of 63·5 cm and a wall thickness of 6·35 cm. The area of the concentrated load was basically 50 mm × 50 mm, but other test areas were also used.

CURRENT TRENDS IN DESIGN AND CONSTRUCTION OF CONCRETE SHIPS (PROFESSOR GERWICK'S PAPER)

In introducing his paper, Professor Gerwick emphasized that cracking was not necessarily a failure criterion in a concrete vessel, and if there was sufficient bonded mild steel, membrane cracking would be adequately dispersed and could certainly be acceptable under occasional overload.

In the ensuing discussion, two speakers queried the alteration of the block coefficient and the Chairman commented that he was confident that a single-curved shell could be produced with a performance equal to that of a double-curved shell.

In response to another question, Professor Gerwick commented that fibres were useful in conjunction with mild steel and prestressing, particularly in areas of high shear, but that he was not advocating their use alone.

In reply to another question, he commented that concrete appeared to be an ideal material for submarines, where the hull was almost entirely in compression, and he knew of a pilot design for a craft about 60 m long by 10 m dia.

A delegate raised the question of concrete strength and Professor Gerwick said that it was normally of the order of 50–60 N/mm^2 cube strength, but in really big ships there must be advantages in very high strength concrete and he knew of designs prepared using concrete as strong as 100 N/mm^2.

THE ROLE OF THE MARINE SURVEYOR (DR SHARPLES AND MR TAIT'S PAPER)

Mr Tait described the involvement of marine surveyors, particularly in the area of offshore structures, where they cover all aspects of the tow, including the analysis of the critical motion for the structure, based on the non-appearance of critical stresses in the towers. This prompted several questions and Mr Tait explained that the curve was drawn for an actual structure and was for the benefit of the crew on board at the time. The philosophy of the curve was to predict a motion which would produce stresses 70% of those which were critical. The motion of the structure could be altered by a change of ballast, reduction in speed, or an alteration in course, but it was a difficult problem and relied, in the last resort, on the skill of the tug master.

Another question raised the problem of 'squat'; that is, the drop of large vessels in channels not much wider than themselves, which was fairly well understood for ships. Mr Tait replied that there were no established theories, but that the question was the subject of some testing. It was known that the amount of sinking varied with the square of the speed, but that, in general, platform tow speeds were very low indeed. However, the clearances beneath platforms in many towing channels were small so that it was certainly an important problem.

In response to the question of instrumentation during a tow, Mr Tait commented that few previous measurements had been taken, but that the National Engineering Laboratory would fully monitor the next tow from Ardyne Point.

STRUCTURAL APPRAISAL OF CONCRETE SHIPS (MR RAPO'S PAPER)

Mr Rapo commented that there was over a century of experience in the design and analysis of steel ships, so it was almost inevitable that concrete ships should suffer a comparison with previous work, although the problem was somewhat difficult in that it involved the analysis of comparatively thick concrete plate elements.

One questioner commented that the hull of a steel ship was sufficiently elastic to accommodate local damage, but enquired what effect there was on the overall stability of a concrete hull if it should become damaged locally. Mr Rapo replied that the analysis of a particular design of concrete hull with holes of 3·8 m and 7·6 m dia. in various positions showed no cause for concern on the question of overall stability. However, he went on to comment that a proper programme of tests on the impact resistance of prestressed concrete ships was certainly needed.

Another delegate enquired whether the analysis had been restricted to symmetrical loading cases and what account had been taken of lateral loads due, for example, to docking and wave action. In reply, Mr Rapo said that the analysis was based on symmetrical load cases. Some general consideration had been given to anti-symmetrical loading, but no detailed analysis had so far been undertaken. Torsion was not considered to be a problem, because concrete vessels generally had a closed section.

QUALITY CONTROL AND MAINTENANCE OF PRESTRESSED CONCRETE SHIPS (MR BEAUMONT'S PAPER)

Much of the earlier discussion on Mr Beaumont's paper centred on the question of damage, and delegates commented that, although prestressing was important for the service conditions, it had very little influence on the ultimate limit state. Another delegate commented that the most serious problem after damage might be corrosion of the steel tendons for which no non-destructive testing procedure existed, although another speaker did point out that much work had been carried out on non-destructive testing techniques at Harwell. In reply, Mr Beaumont pointed out that the object of repair work was to reinstate the ship to its original condition, and therefore the reinstatement of the prestress was important.

Another delegate asked how long the term 'experimental' would apply to a concrete ship, what would be the criterion for the removal of this term, and why could not the first survey be combined with the normal builder's guarantee survey at twelve months. Mr Beaumont replied that the term experimental was applied merely to indicate that the Society reserved the right to examine such a ship at more frequent intervals. Decisions on the periods concerned would be made by the Committee of Lloyds, but six months was currently thought to be the correct period.

PRACTICAL DESIGN AND CONSTRUCTION CONSIDERATIONS ON THE USE OF PRESTRESSED CONCRETE IN SHIP CONSTRUCTION (MR TOSSWILL'S PAPER)

Mr Tosswill commented that he had written the paper basically as a selling aid to the uninformed—that is, owners, bankers and shipbuilders—because, in his view, the hardest problem must surely be to sell the concept of prestressed concrete ships.

One questioner commented that the currently proposed designs of prestressed concrete LNG tankers seemed very susceptible to roll, which increased the 'boil off'. Also, the weight of the vessel itself appeared very high for existing dry dock capacity. In reply, Mr Tosswill said it was true that concrete vessels were perhaps 2½–3 times the weight of an equivalent steel ship, and this could produce problems in conventional dry docks, although it might be possible to distribute the load in some way. A possible solution might be to keep open the original construction site, which might well not be a ship yard in the conventionally accepted sense. In reply to another questioner, he agreed that the tug and barge system seemed to offer advantages, because the cost of a concrete barge was perhaps 20% below that of a steel barge, and a barge would always outlive the motive power unit.

THE MARKET FOR CONCRETE GAS CARRIERS (MESSRS HAY, JANSKY AND MASCARO'S PAPER)

Mr Hay was asked whether the 10% earning related to the cost of the hull, or of the ship. He confirmed that it was the ship.

Mr Stanford noted that the DYTAM design showed a 15% saving. He said that market research showed a need for 100 cryogenic carriers in

1980. Mr Hay agreed that the 1980s would provide a large market; the basic question was when. There was a big gap between 1980 and 1989.

Mr Ffooks commented that the comparison of past forecasts with past requirements showed peaks and valleys like a Swiss landscape. He queried the need for a prototype, which he though had only a psychological effect. Mr Hay thought that a prototype might not be absolutely essential when only relatively small sums were involved; but his company's experience was that little interest was aroused in the absence of a prototype.

GENOA'S FLOATING PRESTRESSED CONCRETE DOCK (PROFESSOR LEVI AND DR COSTANTINO'S PAPERS)

Professor Levi presented both his own paper and that of Dr Costantino. Mr Newman asked whether the density of 2·05 was a dry density; whether additives were included in the mix; how placing and compaction were achieved; and whether the concrete was pumped. Professor Levi replied that 2·05 was the saturated density; no additives were used; the concrete was vibrated; and transport was by conveyor.

GENERAL

Mr Bogdanović described a repair made by Christiani and Neilsen on a reinforced concrete hull in 1961. A hole 4 m × 4 m was repaired using high-alumina cement, and before the next morning the ship had sailed. He deprecated the comparison of new designs with old vessels, in that old drawbacks could easily be attributed to new proposals. This was the more serious in that full records of old constructions were not available.

Both Mr Morgan and Mr Carvill stated that this was not so; full documentation and data were available for many old hulls. Mr Carvill agreed that these often upheld Mr Bogdanović's main point, in that older hulls were built during emergency conditions, when irrational objections by the military prevented adequate research and design. They were built under crash programmes, reluctantly entered into, with inadequate supervision and unskilled labour.

Dr Manning noted the different approaches adopted by naval architects and engineers towards statistical loading conditions, and the very large number of load cases (up to 32) which this had sometime involved. On the question of surveys he noted that the one vital time when inspection was essential was when the concrete was placed. He commented also that rust on tendon ducts was not evidence of tendon corrosion, and asked whether separate special samples might be used for inspection. Lastly, he raised the question of the value of a composite design, using a midship section of prestressed concrete, with steel bow and stern sections.

Mr Rapo replied that the loading combinations to which Dr Manning referred had indeed been conservative, and would not be used in the future. Simple expressions for initial design movements had been included in his paper, and load combinations were considered at a later stage of the analysis. Professor Levi noted that possible loading combinations and their probabilities had been carefully studied, and had then been excluded in the design of the Genoa dock.

Mr Beaumont agreed entirely with Dr Manning on the need for inspection during placing, and for close control of construction. The value of special or sacrificial samples was still hypothetical, but this idea was attractive.

Mr Tosswill thought that a steel/concrete vessel was something of a halfway house; it might help in persuading a shipyard to become involved in concrete construction. Professor Gerwick believed the bow section would be better in concrete, though the stern might be of steel. Mr Tosswill commented that a steel pusher tug and concrete barge was essentially the same concept.

Mr Derrington commented on the large differences of approach between naval architects and classifying authorities, on the one hand, and the concrete industry on the other, and asked for suggestions on how this might be overcome. The panel agreed that the gap was a real one; Mr Tait said that the successful offshore industry pointed the way ahead. Professor Gerwick suggested that naval architects and concrete technologists should take the lead in collaboration. Mr Rapo noted that collaboration takes time to develop; it comes with experience of each other's skills, and requires a flexible approach from all parties. Mr Beaumont placed communication as one of the most difficult problems impeding collaboration.

Mr Sandberg asked whether present-day designs made sufficient allowance for testing. Mr Beaumont thought they did not yet do so, and that there was room for better methods of non-destructive testing. Mr A. B. Anderson pointed out that the maintenance of a construction programme was the responsibility of the contractor, who could not be expected to welcome unnecessary delays imposed from outside. Discouragement of this kind might prevent the adoption of concrete as a hull material.

Mr Furnes commented that Det Norske Veritas used a different design philosophy from that of Lloyds; the latter employed a permissible-stress basis, while DNV used limit-state procedures. In their work on concrete vessels, DNV employed civil engineers; they did not attempt to re-train marine engineers. They also had in hand a project to determine the resistance of prestressed concrete to fatigue and hydrostatic pressure.

Mr Stanford thought that the prime need for the immediate future was the achievement of co-ordination and co-operation between the interested parties, and hoped that the momentum imparted by the symposium would not be damped out. Professor Harris, in his closing remarks, referred to the problem of communication, and thought that the most useful feature of the symposium was in helping to overcome this.

REFERENCES

1. *Proceedings of the conference on concrete ships and floating structures.* University Extension, University of California, Berkeley, 1976.